How to Think
about Information

How to
Think about
Information

DAN SCHILLER

UNIVERSITY OF ILLINOIS PRESS

Urbana and Chicago

© 2007 by Dan Schiller

Manufactured in the United States of America

C 5 4 3 2 1

∞ This book is printed on acid-free paper.

Library of Congress Cataloging-in-Publication Data
Schiller, Dan, 1951–
How to think about information / Dan Schiller.
p. cm.
Includes bibliographical references and index.
ISBN-13: 978-0-252-03132-8 (cloth : alk. paper)
ISBN-10: 0-252-03132-6 (cloth : alk. paper)
1. Information—Economic aspects.
2. Telecommunication—Economic aspects. I. Title.
HC79.I55S35   2007
303.48'33—dc22   2006011275

For Zach

and

in Memory of

Dr. Aaron Rosenbaum

Contents

Part 3: Poles of Market Growth, or a Deepening Crisis?

Acknowledgments

The spirit of this work, and three of its chapters, germinated in writings undertaken for *Le Monde diplomatique*. Warm thanks to the creators of this great voice of critical reason and, especially, to Serge Halimi and Ignacio Ramonet.

Intellectual nourishment came from an extended community of colleagues at the University of Illinois at Urbana-Champaign. I am especially grateful to Chip Bruce, Noshir Contractor, Leigh Estabrook, Kirk Freudenberg, Bob McChesney, Maria Mastronardi, John Nerone, Jan Pieterse, Boyd Rayward, Christian Sandvig, Linda Smith, Inger Stole, John Unsworth, and Bruce Williams.

Three research assistants provided invaluable aid: Jason Kozlowski, Dr. Dal Yong Jin, and Yu Hong. Graduate students, especially Han Dong, Dr. Kelly Gates, Dr. Marie Leger, John Martirano, Victor Pickard, Ben Scott, and Dan Wright, were generous in sharing their thinking.

Exchanges of different kinds with Andrew Calabrese, Edward Herman, Steve Jackson, Rick Maxwell, Toby Miller, Vincent Mosco, Zaharom Nain, Manjunath Pendakur, Lorna Peterson, Greig de Peuter, Ellen Seiter, Lora Taub, Pradip Thomas, and Janet Wasko were indispensable. Yuezhi Zhao's suggestions and advice clarified basic issues and helped gain for this book a wider readership.

It is a pleasure to thank Dr. Willis Regier, director of the University of Illinois Press, for the generosity and care with which he has tended this project.

Family members strengthened the work of writing, as always. Thanks to Susan Davis for inspiration and supportive criticism. Thanks to Lucy Schiller and Ethan Schiller for making room for me to finish and for giving me so many important things to think about. Thanks to Anita Schiller for comments on portions of this text and for bibliographic suggestions. Thanks to Zach Schiller for news clippings and for a bottomless well of enlightening and sometimes hilarious stories.

Scholarly research presentations resulted in valuable clarifications, even as they also helped me to relate this work to the wider world. I am grateful for invitations proffered by Andrew Calabrese, Zhenzhi Guo, Max Dueñas-Guzman, Jyotsna Kapur, Melanie Kimball, Raffaele Laudani, and Lora Taub to present research at the University of Colorado at Boulder, the Beijing Broadcasting Institute, the Central University of Tijuana, Southern Illinois University, SUNY Buffalo, the University of Siena, and Muhlenberg College. Valuable also were presentations to the University of Illinois Institute of Communication Research Seminar on Communications, Culture, and Policy; and the University of Illinois Program on Arms Control, Disarmament, and International Security. Appearances as a guest on Bob McChesney's WILL-based radio show, "Media Matters," helped crystallize my views about telecommunications.

Chapter 1 is a slightly revised version of "How to Think about Information," in *The Political Economy of Information,* ed. Vincent Mosco and Janet Wasko (Madison: University of Wisconsin Press, 1988), 27–43.

Acknowledgment is made to *Critical Studies in Mass Communication* for permission to use material from "From Culture to Information and Back Again: Commoditization as a Route to Knowledge," *Critical Studies in Mass Communication* 11.1 (March 1994): 93–115. See www.tandf.co.uk/journals.

Chapter 3 draws on "Digital Capitalism 2001: The Corporate Commonwealth of Information," in *A Companion to Media Studies,* ed. Angharad N. Valdivia (Oxford: Blackwell Publishing, 2003), 137–56.

Acknowledgment is made to Hampton Press for permission to use a modified version of "Business Users and the Internet in Historical Perspective," in *The Media and Marketization,* ed. Graham Murdock and Janet Wasko (Cresskill, N.J.: Hampton Press, forthcoming).

Chapter 5 draws on "Télécommunications, les échecs d'une révolution," *Le Monde diplomatique* 50.592 (July 2003): 28–29; and "The Telecom Crisis," *Dissent* (Winter 2003): 66–70.

Chapter 7 draws on "Des parasites dans notre quotidien," *Le Monde diplomatique* 48.566 (May 2001): 12–13.

Chapter 8 draws on "Esclaves du portable," *Le Monde diplomatique* 52.611 (February 2005): 1, 22–23.

Acknowledgment is made to Sage Publications for permission to use material published in "Poles of Market Growth? Open Questions about China, Information, and the World Economy," *Global Media and Communications* 1.1 (2005): 79–103.

Preface

Through a linked series of theoretical, historical, and contemporary studies, *How to Think about Information* engages transformative political-economic changes occurring throughout the realm of information technology and culture.

What justification is there for such a project? Is it not passé? "Information," after all, was such a nineties thing, and the Internet bubble is but a fading memory. A glance at the *New York Times* confirms that the subject has diminished in significance; in 2005, the nation's journal of record ceased titling its Monday business section "The Information Industries." Official discourse has evidently moved on.

Interestingly, however, information has not merely been forgotten but explicitly rejected. Where pundits a decade ago insisted that information was the new fulcrum of corporate strategy, by the mid-2000s they are emphasizing with equal certitude that such claims were misplaced. Nicholas G. Carr, the former executive editor of the *Harvard Business Review,* inquired: *Does IT Matter?*[1] His answer is a resounding, "Not any more." From the other direction, the Left, a strangely congruent revision has materialized. *After the New Economy* bids emphatic good riddance to the seeming pipe dreams offered by adherents of the so-called information society.[2]

Skepticism is warranted, and revisions of accepted wisdom are needed; this book contains both. However, I contend that, just as the earlier embrace of information was facile, its dismissal has been too hasty. If we wish to understand the motive force and direction of contemporary society, it is still necessary to think about information carefully and in an extended way. The status of information continues to be elusive because it was never adequately apprehended, either by proponents of information-society theory or their mainstream critics. Questions about information's contemporary structural role therefore remain as significant, and problematic, as they did before the deflation of the Internet bubble.

I argue that, for definite reasons, political and economic elites assigned mounting strategic importance to information beginning around 1970. This was nothing less than an endeavor to renew the encompassing process of market expansion by generating around information a new and expansionary pole of growth for capitalism. If information's significance for this project has dimmed—which is debatable—we should try to learn why. And if it has not, what then? In either case, changes occurring around information have worked their way deep into the economy and society, and these changes demand to be surveyed.

By now it should be apparent that, in this work, "information" operates as a kind of shorthand to include the converging fields of culture, media, and telecommunications. These areas are not identical. In the pages that follow, I untangle some of the semantic intricacies that bind and separate "culture" and "information," and I detail aspects of the process of convergence that is bringing them together. I argue throughout that this inclusive perspective on information holds out the prospect of crucial insights.

We are living through a transition into an informationalized capitalism. As a consequence of policies adopted beginning around 1970, a long-standing process of commodification—the growth of wage labor to produce commodities and of market mechanisms to distribute them—has gripped the global information and communications sector. At the same time, communications and information have come to infuse the more encompassing process of capitalist development.

Basic questions must be asked about this process. What provoked this historical phase-change? How are culture, communications, and information being restructured and repositioned, and how have these changes impacted the overall political economy? What are the political ramifications of the emergent informational emphasis? Will information succeed in rescuing the market economy from stagnation?

How to Think about Information offers, in part 1, an initial explication of the complex process of commodification that now weaves through the informational and cultural realm. Chapter 1 frames contemporary thinking about information in the context of political economy. In chapter 2, arguments about information are further specified and situated historically;

here it becomes important to examine the overlaps and differences between "information" and "culture." Chapter 3 extends the analysis by locating the contemporary political and economic importance of information specifically in the accelerated commodification that commenced around 1970. Accelerated commodification unleashed dramatic policy changes in intellectual property law, ownership and control of information resources, and telecommunications-system development to sustain rapid growth of private property in information. It also helped deepen an already worrisome antidemocratic tendency.

In part 2, I analyze the commodification process in several specific contexts. Chapters 4 and 5 focus on telecommunications networks and services, to lay bare how these "critical infrastructures" have been altered and to what effect. Established opinion notwithstanding, this history cannot be read solely from the supply side—that is, as a straight-line function of the telecommunications carriers' strategy and structure. Nor has the logic of network reorganization been benevolent or free of conflict. After detailing the neglected but pivotal role played by business users in telecommunications-system development in chapter 4, I show in chapter 5 that the liberalized policies they demanded led to a damaging political-economic crisis.

Chapters 6 and 7 redirect attention to the more widely familiar domain of culture. Through an extended appraisal of the rapidly changing culture industry, transformative developments in technology and political-economic structure are assessed. Accelerated commodification in this sphere has engendered not only growing "convergence," which erodes divisions between media long separated as a matter of deliberate policy, but also a transnationalizing capital logic. Far from deviating from these trends, Internet systems and services incarnate them. New directions of growth in the sponsor system, the culture industry's long-established institutional patron, are then canvassed. Chapter 8 takes up the explosive growth of wireless technology. I show that, drawing alike on necessity and desire, wireless extends a long-standing and often disabling trend in our social organization that Raymond Williams calls "mobile privatization."

In chapter 9, many of these themes are united in a different context, through analysis of the reintegration of China into the world market system. Examining how the most dynamic region of the world economy has combined with its most dynamic industry sector allows us to return to an overarching question: Has a solution been found for one of capitalism's most deep-seated and intractable problems, its continuing tendency toward economic stagnation as a result of overproduction?

This book, which synthesizes work produced over a period of twenty years, engages processes that continue to be alive, incomplete, and fraught with conflicts and contradictions. I have simply attempted to frame some of the key issues and to sketch some provisional answers. For better or worse, how to think about information remains in vital ways an open question.

How to Think
about Information

PART 1

The Roots of Informationalized Capitalism

1

How to Think
about Information

In 1982, the American Express company heralded its elevation into the select group of thirty corporations that comprise the Dow Jones Industrial Index as follows:

> Our product is information—information that charges airline tickets, hotel rooms, dining out, the newest fashions, and even figures mailing costs for a travel magazine; information that grows money funds, buys and sells equities, and manages mergers; information that pays life insurance annuities, figures pricing for collision coverage, and creates and pays mortgages; information that schedules entertainment on cable television and electronically guards houses; information that changes kroners into guilders, figures tax rates in Bermuda, and helps put financing together for the ebb and flow of world trade.[1]

American Express was not unique. Companies engaged in making and selling entertainment, banking, communications, data processing, engineering, advertising, law, and other information-intensive services have played an increasingly critical role in overall U.S. investment, employment, and international trade. Major manufacturers such as General Motors, McDonnell Douglas, and General Electric were diversifying into information either through in-house activities, commercial provision, or both. An administrative discipline known as "information resource management" was finding broad application across the business world. This discipline treats information "as a resource like other resources such as money, personnel, and property, which have values and costs and are used in achieving program goals."[2]

In short, as one authority declared, we "can no longer deny that information is becoming a commodity."[3]

Not just any commodity, either, but a fundamental source of growth for the market system as a whole: information, many proclaimed, had become the essential site of market growth. Yet the source and true character of this general shift in the social definition of information remain obscure. Where is the common ground shared by a college course in anthropology and computerized credit data, or between a newspaper account of a city council meeting and genetically coded messages? How, when, and why does information become economically valuable?

Throughout the postwar era, many writers have grappled with such questions. At least three broad, often overlapping conceptual approaches to the subject have emerged. In spite of serious flaws, each tradition may be utilized critically as a means to explore how to think about information.

Informational Theories

As late as 1933, the *Oxford English Dictionary* gave no hint of the profound shifts beginning to occur in the conceptualization of information. The dictionary revealed only that "information" had been in currency in English since Chaucer, when it denoted an item of training or an instruction, and that the word then accrued several additional meanings: an idea, the communication of news, or a complaint against a person presented in court.[4] By the end of the 1940s, however, mathematicians and engineers had forced a radical break with these past usages.[5] Their development of statistical formulae for measuring the amount of "information" within a system proved of immediate utility to communication engineers trying to design cost-efficient transmission channels of appropriate bandwidth, or information-carrying capacity.[6] Yet the "mathematical theory of communication" depended upon a dramatic redefinition that formalized the ongoing transformation of "information" into an encompassing category of apparently sweeping relevance and explanatory potential.[7] Information, these theorists asserted, is a measure of organization, pattern, structure, or—in Klaus Krippendorff's more recent treatment—of a potential for organizational work.[8] "[T]here is a widespread feeling," wrote one scientist receptive to information theory's universalistic ambition, "that information theory is basic to a thoroughgoing consideration of all organized systems."[9]

"Information theory" promised to unlock the inner workings of diverse systems—collections of related entities—"from steam engines to human societies," as one of the eminent participants in a conference underwritten by the Macy Foundation put it.[10] In 1950, for example, an article in *American Scientist* declared that "consideration of the effects of information storage and information transfer on physical, chemical, biological, psychological, and sociological systems" might "help in understanding and predicting many of the aspects of our universe."[11] Early confirmation of the value of this approach appeared evident in a contemporary breakthrough in biology: the discovery of the precise sequence of the DNA molecule, which forms "the code which carries the genetical information."[12] A hitherto undetected but potentially vital informational component of physical, chemical, biological, and even social systems could be sought after and, it was believed, explicitly specified.[13]

However, despite significant refinement and conceptual augmentation (such as Ludwig von Bertalanffy's "open systems" concept), information theory encountered difficulties when applied to social processes. These difficulties were due largely to a tendency, common then and still evident today, to operationalize the system concept. Systems, it was held, require rigorous codification of input and output "variables," their "values," and above all, their behavioral relationships. This process of operationalization tends to impose a formal, mechanistic order on the contingent, conditional, and often unclearly interrelated historical realm of human social agency.

Compared with telephone networks and even biological organisms, societies appeared to be "exceedingly complex systems," whole chunks of which had, of necessity, been relegated by theorists to "black boxes" whose internal workings "cannot be comprehended" but should be explained at some point further along.[14] This point, however, never arrived. Even more damaging, any possible social determinants of information tended to drop out of the analysis. "[W]ithout materials there is nothing, and without energy nothing happens," one borrower from this tradition wrote. "But without information, nothing has meaning: materials are formless, motion is aimless."[15] Are there really no irreducibly social agencies and historical relationships conditioning the organization and use of information, energy, and matter? Information theorists implicitly identified the essential structuring agency of all systems in information itself. They not only sought an informational component of organized systems but also hoped to develop a unique informational plane of analysis to explain its operative features. Therefore they sidestepped the possibility that information—at least in social "systems"—might be a product of social institutions. Though matters of indifference to information theorists, the following questions are critically relevant: What social forces structure information? How have they developed? Over what range of "systems" do they operate?

A quite different school of thought did attempt to elucidate a social framework for information. By the late 1960s and early 1970s, theorists of an emerging "postindustrial" or, later, "information" society became prominent. Postindustrialists argued that new "intellectual technologies"—above all, the computer—were dramatically discontinuous with earlier systems of information processing and control. The new technologies would be as central to the emerging society as "machine technology" had been to its industrial predecessor. Postindustrialists often claimed to find a "new class" of office employees—"knowledge professionals" or "information workers"—constituting the preponderant segment of the employed workforce in the developed market economies. Finally, they argued that information itself had become the transforming resource of social organization. For Daniel Bell and those generally lesser thinkers who followed him, the postindustrial society broke with and transcended the elemental relations— including, most crucially, the opposition between capital and labor—that had shaped its antecedent.[16] Knowledge was supplanting capital and labor as the decisive factor of production.

There were ideological advantages to declaring that postindustrial society constitutes a radical break with the past. By emphasizing discontinuity, however, it was the rather dismal visage of the present that was obscured. The selective silences in the writings of the postindustrialists and their successors are notable: about the crisis of empire occasioned by the devastating political and economic impacts of the U.S. war in Vietnam; about the economic slowdown—to which intensifying international competition contributed and responded—that threatened to undercut the brief "American century"; and about the unprecedented transnationalization of the entire market system. Rather than attempting to contextualize the important changes in the information sector that they did identify within these determining historical circumstances, the postindustrialists chose to abstract from them. Information society analysts such as Harlan Cleveland carried this tendency to even greater lengths. Ironically, Cleveland even criticized Daniel Bell for his coinage, the "postindustrial society," on the grounds that the term put *too little* distance between a promising future and a rusting past: "Can't we find a term for the future that goes beyond saying it comes after the past? Surely postindustrial is too reductionist a tag for so different and exciting a prospect, and too economic a name for a period in which the discoveries of science, the innovations of technology, and integrative thinking about politics, culture, and psychology will be at least as important as economic analysis to an understanding of what's going on."[17]

Yet an even more basic conceptual flaw is evident throughout the writings of the postindustrialists. They commonly pinpoint the source of information's economic value in supposedly intrinsic attributes of information itself. "The information resource," claimed Cleveland, "in short, is different in kind from other resources"—as if, in finding that a shoe is not a table, he had somehow fixed at last on the former's essential economic nature.[18] Not subject to the laws of thermodynamics, information is "expandable, compressible, substitutable, transportable, leaky, shareable." These "inherent characteristics," Cleveland declared, divulge the vital clue to information's mounting economic importance.[19] Such claims grew wearisomely frequent within mainstream economic arguments, as did the warning that because information is a public good,[20] "markets for information products may not operate in the same ways as markets for tangible commodities."[21]

This reasoning resurrected a long-standing economic fallacy. As Rudolph Hilferding pointed out at the beginning of the twentieth century, such a theory of economic value invalidly relies upon categories that "are natural and eternal entities."[22] It substitutes for the historical development of social relations among persons the purportedly immanent qualities of things. Why was the status of information not a major topic of economic theory in 1700, 1800, or 1900? Why was it only after World War II that the economic role and value of information took on palpable importance? The advocates of information's innately distinctive economic role were necessarily unconcerned with such questions. They found it difficult to explain the history of their subject without retreating into technological determinism: the "computer revolution" thus was charged with responsibility for the unprecedented visibility and economic significance of information. But this is not a satisfactory answer. Why was there a "computer revolution"? Why only in the postwar era? Why predominantly in the developed market economies? And what kind of an upheaval did this "revolution" actually portend?

We will not comprehend why and how information becomes economically valuable by beginning from its supposedly intrinsic attributes; we cannot uncover its real social framework in this fashion. What if, however, we suppose that information is not inherently valuable? What if only a profound social reorganization can permit information to become valuable? What sort of historical changes would be required for such a sweeping and dramatic revaluation?

To answer these questions we will introduce a key distinction between information as a resource and information as a commodity. Using this elemental distinction, we may begin to grasp the nature of information in contemporary society.

A resource is something of actual or potential use. That is all. The soil, the sea, and the spectrum are resources. But all resources are not commodities. Only under particular conditions can they be transformed into commodities. A resource is anything of use, anytime, anywhere, to anyone; but a commodity bears the stamp of society and of history in its very core.

Lacking this distinction, information-society theorists pursue an erroneous consideration of how information is innately different from other resources. Employing it, however, permits us to consider how information is socially identical to other commodities. When we begin to study the production and use of information resources within history, we find that they have experienced the same series of changes in social organization as other resources claimed by capitalism and transformed into commodities: *all are produced increasingly by wage labor within and for a market.* Oblivious to this social transformation, information-society analysts evince an implicit willingness to treat all resources, including information, as commodities. With the commodification of information in mind, however, we enter the domain of political economy, which seeks to comprehend the historical evolution of the market system itself.

It might have been assumed that political economy would be at the forefront of the social analysis of information. Yet the third major way of thinking about information, stemming from political economy, paradoxically dismisses an economic role for information altogether. This line of reasoning holds that activities such as advertising, market research, law, financial services, and other information-intensive pursuits are simply not productive. As with the two previous schools of thought, this theoretical tradition yields useful insights upon investigation.

Advocates of this view assert that informational functions such as advertising are unnecessary—even harmful—and that therefore they should not be treated in the same fashion as such clearly beneficial pursuits as farming, bauxite mining, steel smelting, and automobile assembly. Prominent exponents of this position are Paul Baran and Paul Sweezy, who, after arguing trenchantly that advertising "constitutes as much an integral part of the system as the giant corporation itself," go on to claim that, nonetheless, advertising expenses "are manifestly unrelated to necessary costs of production and distribution—however broadly defined."[23]

Baran and Sweezy justify their argument on moral and theoretical grounds.

They echo a long-standing concern with differentiating and supporting "useful" against "unuseful" social labor. Such a distinction was essential for nineteenth-century journeyman artisans who contrasted their own productive crafts with the connivings of "monopolists" and "speculators." The latter, personified by bankers and lawyers, were hastening the expansion and rigidification of wage relationships.[24] But the distinction between producers and nonproducers has persisted down to the present as a widespread mistrust of "parasitic" paper-pushers and other members of information-intensive occupations.

A theoretical argument seems to lend this suspicion credibility. When reduced to essentials, the argument runs as follows: no matter how important to the functioning of the market economy, virtually all information-intensive employments fall within the sphere of circulation of capital and not of production. Once assembled, a car must be advertised, marketed, and financed, but these functions are ancillary to the production of the automobile itself. They add no new value in their own right. To the older strata of "surplus eaters"—unproductive participants in the economy—are added a host of new ones: "corporate and government bureaucrats, bankers and lawyers, advertising copy writers and public relations experts, stockbrokers and insurance agents, realtors and morticians, and so on and on seemingly without limit."[25] All are part of a vast social wastage created by the need to dispose of a mushrooming economic surplus under conditions of monopoly capital.

This logic, however, is flawed on three counts. First, the occupations consigned by Baran and Sweezy to the category of circulation clearly do not constitute the entire information sector. Scientific research and engineering spring to mind in this context; and what about the differently situated labor of teachers and librarians? The information-intensive occupations are simply mischaracterized if they are said to function solely within the domain of circulation. Second, the distance between circulation and production has narrowed dramatically. Advertising, market research, industrial research, and development have grown ever more tightly coupled to production and distribution, particularly in the consumer goods and services sector. Toothpastes, detergents, breakfast cereals, and video games, for instance, are not produced until exhaustive research has identified which product features, market locations, and demographic groups should be targeted to facilitate maximum profits. It is therefore increasingly impossible to disentangle "circulation" from "production." Finally, the occupations consigned to circulation by these analysts have changed fundamentally since the mid-nineteenth century, when the dichotomy between productive and unproductive labor played a more valid role. This crucial point requires amplification. *"For labour to be designated productive,"* wrote Karl Marx,

> qualities are required which are utterly unconnected with the *specific content* of the labour, with its particular utility or the use-value in which it is objectified. Hence labour with *the same content* can be either productive or unproductive.
>
> For instance, Milton, who wrote *Paradise Lost,* was an unproductive worker. On the other hand, a writer who turns out work for his publisher in factory style is a productive worker. Milton produced his *Paradise Lost* as a silkworm produces silk, as the activation of *his own* nature. He later sold his product for £5 and thus became a merchant. But the literary proletarian of Leipzig who produces books, such as compendia on political economy, at the behest of his publisher is pretty nearly a productive worker since his production is taken over by capital and only occurs in order to increase it. A singer who sings like a bird is an unproductive worker. If she sells her song for money, she is to that extent a wage-labourer or merchant. But if the same singer is engaged by an entrepreneur who makes her sing to make money, then she becomes a productive worker, since she *produces* capital directly. A schoolmaster who instructs others is not a productive worker. But a schoolmaster who works for wages in an institution along with others, using his own labour to increase the money of the entrepreneur who owns the knowledge-mongering institution, is a productive worker.[26]

Labor is productive, in other words, if it creates a surplus for a capitalist over and above the wealth that it consumes in order to be capable of producing at all.[27] No matter how repellent the function of a given kind of labor, it is productive if it "is taken over by capital" so as to contribute to accumulation by means of the wage relationship and market exchange.

Toward the end of the nineteenth century, the majority of the information-intensive activities dismissed as unproductive by Baran and Sweezy were performed, if at all, by self-employed individuals. Law, accounting, advertising, research and development, and so forth were neither organized as autonomous capitalist enterprises nor yet assimilated within the division of labor of the giant firm. They were carried out almost exclusively by individuals or partners brokering their services to capital in exchange for money; no capitalist directly appropriated their own surplus, and thus they did not contribute to capital's productive self-expansion. For this reason Marx dismisses them as "peripheral phenomena that can be ignored when considering capitalist production as a whole."[28]

Information services were not, however, to be left permanently to the realm of self-employment.[29] Marx left open the possibility that the definition of productive labor could broaden with the historically expanding arena of capitalist production: "[F]or the most part, work of this sort has scarcely reached the stage of being subsumed even formally under capital, and belongs essentially to a transitional stage."[30]

Today, these once "peripheral phenomena" have become profoundly important. Complex historical processes have empowered capital to impose wage relationships on many formerly exempt categories of social labor. Technical innovation, including innovations in the areas of information storage, processing, replication, and distribution, hastened the separation across a growing range of employments of the producer from the product and the process of production. The expanding scale of capitalist enterprise demanded and made possible continuing innovation in information technology and a colossal expansion of the gamut of business services performed by Baran and Sweezy's "surplus eaters." These, however, were not surplus eaters at all. Drawn into the giant firm or entering the employ of specialized service corporations, formerly self-employed professionals were instead increasingly transformed into directly productive wage laborers. Their labor not only contributed to the value of the automobile, to put it another way, but also, more importantly, produced a surplus for the owners of accounting firms, advertising agencies, research and development laboratories, public relations companies, and corporate marketing departments. This trend is continuing—indeed, accelerating—across the fields of education, librarianship, and medicine, among other areas of social labor being taken over by capital.

In contrast, Baran and Sweezy abstract from the elemental and historically dynamic relationship that is capitalism—the wage relation—in favor of a Sisyphean attempt to distinguish productive from unproductive labor in terms of a hypostasized set of productive activities. They first confuse socially useful or desirable activity with economically productive labor and then try to determine whether a given pursuit is productive on the basis of its contribution to an arbitrarily truncated "basic" sector of the economy. Their argument is thus ahistorical and comes perilously near to advancing a modern variant of the fallacy introduced by the economists known as the Physiocrats in the late eighteenth century: that only labor that works on the land, in agriculture, is productive. In its modern form, the argument stresses not only agriculture but also mining and manufacturing; it may even allow that the zone of productive labor comprises the "material" production of goods per se. In either case, it is only by fiat that information-intensive (and a range of other) labor is consigned to the unproductive realm. Such a view takes insufficient account of the fact that capitalism, in its essence a dynamic form of social organization forever dependent on identifying and exploiting new areas of social labor, has moved from agriculture to manufacturing and beyond over the course of its history.

Such an argument, finally, also engenders a profoundly misleading approach to the changes unfolding in the contemporary political economy. To explicate this point, it is helpful to examine an analysis of the so-called strange recovery of 1983–84 by the editors of the *Monthly Review*.[31] Paul Sweezy and Harry Magdoff's article presents data on investment in producers' durable goods—the machinery, equipment, and transportation apparatus as well as structures whose use undergirds the entire U.S. economy. The data showed that investment in "high-tech" products, including computing, accounting, and office machines, communication equipment, and instruments, far outstripped investment in traditional agricultural and manufacturing equipment over the 1979–84 period.

This portrait of capital investment trends cannot be challenged. By 1985, in aggregate (1982 dollars), corporate outlays for such high-tech products made up an astounding 36 percent of nonresidential purchases of producers' durable goods; this proportion had increased steadily since the 1940s but accelerated markedly in the wake of the 1974–75 recession.[32] Electronic products moved ahead of factory machinery and mobile equipment to become the largest single category of capital equipment spending in 1985—which they have remained.[33] The question is, What to make of these striking facts?

Sweezy and Magdoff view this spectacular change in the overall character of capital investment as "for the most part concerned with making money rather than with making goods."[34] The billions upon billions of dollars invested in computers, telecommunications networks, office equipment, and advanced instrumentation of every kind "have less to do with production than with finance" and merely "sustain and foster speculative fevers in the process of servicing customers, designing strategies for investment and speculation, and inventing new instruments for speculation."[35] In contrast to some of the technological revolutions of the past, which "have served to lift capitalist economies out of a stage of stagnation and to set in motion a long period of rapid growth,"[36] information technology yields scant economic promise. This is because the new electronic and communication technologies ostensibly have had only "minor secondary effects" in the areas of manufacturing and construction, "the basic industries."[37] Far from heralding a new wave of capitalist expansion, information technology in this view betokens only a more serious phase of stagnation giving rise to rampant speculation. Neither the general tendency to chronic economic stagnation nor the specific "speculative fevers" so much in evidence today, both of which the editors of the *Monthly Review* did so much to document and analyze, can be doubted; indeed, we will see in chapter 5 how deeply they have engraved themselves into what I call "informationalized capitalism." However, owing to their a priori assignment of productive labor to one historically antecedent segment of the overall economy—the "basic industries" of agriculture, construction, and manufacturing—Sweezy and Magdoff refuse to recognize the empirically observable and continually escalating contributions of the new information technology across and throughout all economic sectors. This refusal is made simpler and more palatable because the authors treat the U.S. economy strictly in national terms. That the prime unit of economic activity is today the transnational corporation—a fact of which these authors are well aware, having written eloquently about it elsewhere—in this analysis disappears from view. So too does the fact that these transnationals, by far the most important source of demand for the high-tech capital investments described above, are employing these investments precisely to unify, control, and further expand their productive profit-making activities on a global scale.

By the 1980s, satellites were being used to interconnect computerized printing facilities of newspapers such as the *Wall Street Journal, USA Today,* the *Toronto Globe and Mail,* and the *Financial Times,* permitting them to garner advertising revenues in national and transnational markets. Research and development laboratories shared data and sophisticated information-processing capabilities across borders by means of packet-switched computer networks. Book publishers deployed computer systems to facilitate tighter inventory control and to expedite distribution among publishers, wholesalers, retail chains, and libraries. Credit reporting and financial information services migrated to electronic networks to deliver data, "treasury management" programs, and money itself to corporate subscribers. Manufacturers like General Motors and Boeing moved to link engineering and design directly to production and to integrate "islands" of automated machinery through transnational computer-communications systems. Agribusiness combines and energy companies relied on satellite remote sensing to monitor global harvests and to search out oil and mineral deposits. Is this manifold evidence of the integration of information technology into production and distribution simply to be dismissed as economic waste?

And what about the labor of the millions of wage earners who are today engaged, directly and indirectly, in the use of these new information technologies? Utilizing the new productive processes supported by continuing massive capital investments in such technology, these workers create conventional manufactured, mined, and farmed commodities and information products and services. Heavily concentrated in the United States and other developed market economies, information-services employment became one of the fastest-growing segments of the labor force. Employment in computer and data processing services more than tripled between 1974 and 1984; job growth in advertising—a comparative laggard in the area of the economy known as business services, which includes most contract information services—was still more than twice as large as the overall industrial average.[38] Over this same decade, the number of nonsupervisory employees in the entire business services industry doubled, to just over four million, showing a rate of growth four times that of the nonagricultural economy as a whole.[39] And this does not include the additional information-service workers laboring within diversified transnational corporations as in-house employees. Can these millions of wage earners continue to be thought of merely as unproductive laborers?

First in agriculture, then in manufacturing and other areas, capitalist forms of economic organization have continued to take hold of production.[40] To claim otherwise is to reveal an arbitrary blind spot to the fundamental trends within the political economy, centering on the crucial role of communications and information, that Dallas Smythe identified years ago.[41] Although the prospect of a national economy whose role in the emerging transnational division of labor centers on services (and military industries) may be abhorrent, that economy has only extended capitalist commodity production to new spheres. To that extent, it remains economically productive. Indeed, it could be argued that, despite and in vital ways *because of* the tendency to stagnation so evident today, an emerging information industry has acquired exactly those features characteristic of previous leading-edge sectors, such as the railroad, electrical, and automotive industries, each of which spearheaded sustained economic expansion in its era.

Conclusion

Wage labor and markets are always incomplete products of human social organization. Neither bursts suddenly on the scene in finished form; their constant expansion has been a chief task confronted by capital. Wage relations emerge unevenly, often the result of protracted struggles over the character and control of production in particular trades. The development of the market into an increasingly encompassing and integrated mechanism for commodity exchange has also been a slow process. The state, communications and transportation technologies, and an intensifying need to dispose of surpluses have decisively contributed to it. The penetration of capitalism has thus always been and continues to be a dynamic and unfinished process.

It is a process that has two basic dimensions: as new peoples are brought under the wage relationship, new products, services, and productive processes facilitate the self-expansion of capital throughout growing segments of social life. In principle, its endpoint, assuming that neither Armageddon nor basic social transformation intervenes, arrives only when all social labor is waged and all products and services are exchanged as marketed commodities. What is significant today is the "progress" toward this theoretical endpoint. The wage form is far more universal than at any previous time, the impact of automation notwithstanding; over the course of the present century it has been imposed globally, making especially impressive inroads into the so-called less-developed countries and, most spectacularly, in the former Soviet bloc and in China. The profound importance of this expansion was recognized by Alan Greenspan, then chairman of the Federal Reserve, who testified in 2005 that the "addition of these workers plus workers from India—a country which is also currently undergoing a notable increase in its participation in the world trading system—would approximately double the overall supply of labor."[42] Within the developed market economies the wage has also continued to advance, extending lately to embrace formerly independent professionals and an unprecedented proportion of women. Over this century, at the same time, the world market has been woven ever more tightly, to the point where now it is made extraordinarily difficult— politically, economically, and often militarily—for any country or region to extricate itself or to maintain itself after its initial extrication.

This is the historical and analytical matrix within which the contemporary transformation of information must be situated. To the information theorists' claim that information denotes organization, we say yes, but information itself is conditioned and structured by the social institutions and relations in which it is embedded. These social relations are today creating a specific form of capitalist organization across an unprecedented range. To the postindustrialists' assertion that the value of information derives from its inherent attributes as a resource, we counter that its value stems uniquely from its transformation into a commodity—a resource socially revalued and redefined through progressive historical application of wage labor and the market to its production and exchange. The wage has been imposed continually on new fields of social labor, including information. Markets have been developed as the determining exchange mechanism for an ever-expanding spectrum of commodities, again notably including information. For these reasons, finally, as against those political economists who dismiss the information sector as unproductive, we suggest that the information commodity has become the prime site of contemporary expansion—such as it is—within and for the world market system.

2

Culture, Information, and Commodification

This chapter extends and deepens our conceptual engagement with information. Its point of departure lies in the acknowledgment that the study of information is uncomfortably disparate. The term refers to diverse phenomena in library and information science, engineering, communication, sociology, law, economics, literary criticism, history, and other fields. And the contemporary concept of information stems from two distinct traditions of intellectual engagement.

One of these traditions was gestated during the first half of the twentieth century, mainly within the context of engineering complex, high-technology control systems. Although some early developers of what was formalized mathematically as "information theory" questioned whether it held broad applicability,[1] most did not,[2] and influential figures such as Warren Weaver of the Rockefeller Foundation emphasized the theory's multifarious potential. Other academic luminaries likewise celebrated its prospectively sweeping contributions; Wilbur Schramm, for example, led the attempt to import information theory into communication study, declaring, "We have every reason to suspect that a mathematical theory for studying electronic communication systems ought to have some carry-over to human communication systems."[3]

Schramm's optimism was not altogether misplaced. As the historian David A. Mindell shows, information theory did not develop in the asocial context of strictly physical systems.[4] It originated in programs of innovation aimed at controlling the behavior of complex processes, containing physical and human components, such as telephone networks, electrical grids, and naval artillery-fire control systems. A powerful, formal conception of signals was born of these initiatives—a conception within which speech, image, and text began to merge, for engineering purposes, into what could begin to be apprehended as a common electronic information stream. A comparably inclusive ambition converged in attempts to build stored-program electronic computers as (mostly) general-purpose systems. As early as 1948, for example, a promotional brochure declared that the UNIVAC computer—not yet on the market—could be used for "air traffic control, census tabulations, market research studies, insurance records, aerodynamic design, oil prospecting, searching chemical literature, and economic planning."[5]

However, in a key respect, information theory's generalization after World War II is widely seen today to have been facile. The theory was maladroitly transposed from processes of signal transmission to socially grounded exchanges of semantic content—that is, to the meaningful realm of culture. Processes of communication were thereby stripped of any constitutive social significance, so that, as L. David Ritchie explains, "the statistical characteristics of a code" were confounded "with the cognitive and social processes of communication."[6] As Ritchie underscores, "Even the most routine forms of human communication can be understood only in the context of the social relationships in which they take place."[7]

Despite, or perhaps because of, this confusion of "information" and "culture," information theory contributed to the reorientation of the social sciences. During the Great Depression, topical issues of profound importance—the sources of economic stagnation and growth; the role of corporate capital in politics and culture; the functions of the state; the changing status of social science itself—had been engaged. With the onset of the cold war, however, the widening assimilation of information theory played a role in displacing these still-vital concerns. As communication study, for example, bid for legitimacy, information theory was incorporated at the expense of a prior emphasis on mass persuasion—the power of radio, film, and other new media, often in concert with interpersonal networks, in conducting modern propaganda—and on the dominative processes that underlay and infused this function. The field's chief intellectual concerns turned instead toward abstract, formalized discussions of information senders, receivers, and channels, within which attention was focused on measures of attitudinal and behavioral change among individuals and networks of individuals.[8]

That information constitutes an inclusive and perhaps transcendent dimension of varied "systems," however, has widely continued as an article of faith. Information's ostensible field of reference came to encompass physical, psychological, biological, and social phenomena.[9] The definition of "information" was similarly extended as "the ability of a goal-seeking system to decide or control,"[10] or, as Klaus Krippendorff eventually framed it, a potential for organizational work.[11] Conceptions of information (which might remain valid in the context of traffic control but were mere placeholders when applied to socially encoded processes) were rendered in inclusive but essentially formalistic terms: messages, patterns, "the communication of relationships."[12] Such a conception was adopted by the influential theory of "postindustrial society," in which information also came to be accorded decisive significance.

"By information," writes Daniel Bell, "I mean data processing in the broadest sense; the storage, retrieval, and processing of data becomes the essential resource for all economic and social exchanges. These include: data processing of records . . . data processing for scheduling . . . data bases."[13] For Bell and most of the lesser analysts who drew on his conception to promulgate theories of information society, information again is used to encroach upon and to render irrelevant the socially encoded processes of semantic meaning that permeate human experience.

Viewed as "data processing in the broadest sense," "information" might have been seen to broadly overlap with "culture." Instead, "information" simultaneously encompasses and conceals much of what is referenced by the anthropological sense of "culture." Where is the boundary between information as "data processing in the broadest sense" and culture as the realm of patterned meaning? Are computer programs not cultural artifacts? Do expressive forms not possess an informational potential for organizational work?

As Bell developed postindustrial theory, the concept of culture was also undergoing a complex process of revision. On one side, the humanistic connotations of "culture" skewed in a different direction than the scientific accents of postindustrialism. On the other side, more direct difficulties were posed by an ongoing reorientation of "culture" toward the conflicts and struggles that shape ordinary human experience.[14] In the United States, the civil rights movement, the anti–Vietnam War struggle, and the 1960s counterculture indicated that conflictive elements figure deeply in culture. From the global South, Mao's Cultural Revolution and the rising antagonism to "cultural imperialism" comprised additional signs that "culture" no longer could be used to denote an apparently unbroken traditionalism. Such usages moved actively against postindustrialism's benign prophecy of progress.[15]

Bell himself consistently accepts that "culture" signifies the expressive symbolization of experience. But between this and what he sometimes terms the techno-economic order, he postulates an unbridgeable gap. Bell sought to portray culture, politics, and the economy as disjunctive domains, existing on separate planes and operating on mutually independent and sometimes even antagonistic principles.[16]

"Information," in this scheme, operates discursively to establish a vital and unremarked distance from experience and consciousness. Its scientist aura of objectivity lends itself to overarching reifications; the absence of any clear distinction between the terms "information society" and "information economy" is telling. "Information" thus directed analysts away from "culture's" rich debates—employing references to "high culture," "mass culture," and "popular culture"—over whose experience could be normative and whose should be made general. In these ways, "information" abstracted from social life; it requires a contextualizing noun—"information society"—to expand its field of reference in this direction. "Culture," in contrast, generalizes on its own to a whole, though perhaps problematically shared, way of life. In shifting discussion onto what they presented as new terrain, the ideologists of "information" thus created analytical distance from lived, often socially conflicted, experience.

Despite their unobserved convergence, it remains crucial that the two concepts remain distinct. If "information" constructed a scientistic fantasy, it also instigated and rewarded a search for new conceptual entry-points into the social process: the reorganization of work within modern industry, shifts in the international division of labor, and the history and applications of information technology. Most important, perhaps, for critical analysts, "information" enabled a resurgence of analysis and debate over the encompassing principles and practices of societal evolution and over the place of a changing accumulation process within this larger history.[17]

While the intellectual extensions occurring around new usages of "information" must not be abandoned, neither must "culture" as conflicted human experience be traded away. Rather than searching for a mechanical convergence between "culture" and "information," we must try to make more explicit the problematic links between information as the potential for organizational work and culture as the experience of capitalist society's divided population.

The Commodification Process

We may begin by thinking afresh about information as a commodity in an attempt to constitute a more satisfactory approach than ensconced postindustrial theories permit. Whereas these theories begin with industrial society, which they then claim is being transcended, information-commodity theorists begin with capitalism, which they argue is not being transcended. Postindustrial analysts build their work on the premise that information is intrinsically anomalous and informational labor innately dichotomous with other forms of work. Information-commodity theorists assume instead that information can be compared usefully with the vast range of other commodities whose production depends on common capitalist relations of production.

What is a commodity? This crucial concept allows historically patterned and conflicted social experience to be reintroduced into the discussion. In chapter 1, I argued for an explicit, restricted definition of "commodity," one lodged specifically in social relations rather than, for example, individual preferences and predilections. A commodity is not merely a product or a resource, a thing of use to anyone, anytime, anywhere; it also may be distinguished from another current usage: a staple or mass-produced (as opposed to a custom-made or handcrafted) product.[18] A commodity is a resource that is produced for the market by wage labor. Whether a tangible good or an evanescent service, universally enticing or widely reviled, a consumer product or a producer's good, a commodity contains defining linkages to capitalist production and, secondarily, to market exchange.

In the present context it is helpful to focus not on the commodity in itself but rather on the commodification process.[19] An uneven but ongoing process of commodification is foundational to capitalist development; its historical generalization throughout the informational sphere constitutes a landmark of the contemporary political economy. With regard to information, commodification covers two cases or aspects. First are those instances where information is the final product; second are those in which information is an intermediate component of production. In either case, as various analysts have suggested, we should scrutinize especially the means whereby capitalist social relations are insinuated or accepted into what had earlier been noncapitalist forms.[20] This encompasses interrelated and often conflicted historical tendencies toward production by wage labor and private appropriation; exchange via capitalist markets; creation of new means of producing and distributing information and cultural commodities; attendant restructurings of the labor process; and widening acquiescence to the idea that particular genres of information should have not only costs but prices (to take a contemporary example, cable television supplants "free-to-air" broadcasting with subscription television services and pay-per-view programs).

Again, there are no intrinsic, transhistorical reasons why any particular field of labor should unfold in this direction at a particular time, let alone end there. The social field is structured by other, equally important processes that often intertwine with commodification in complex ways. It is important to underline that specific conditions and pressures exerted by a capitalist political economy, forever requiring new markets, new materials and production processes, and new and cheaper sources of appropriately skilled or deskilled labor, engender the commodification of particular practices of information and culture production. A four-cell diagram indicates the possibilities, together with illustrative examples:

Wage and Market	Wage Non-Market
TV Cameraperson	Federal Census Statistician
Unwaged Market	Unwaged Non-Market
Bestseller Writer	Bedtime Storyteller

The commodification process is not best studied within any particular occupation. Usually it is appropriate to think in terms of the range of labor processes and interlinked industries needed to produce and distribute a particular commodity.[21] A book thus becomes a commodity not necessarily because its author works directly within a wage relationship; usually he or she does not. The book becomes a commodity during the production process because, within the overall organization of the publishing industry, wage labor and market exchange are the norm. This does not mean that all segments of the division of labor in publishing must be bound up in wage relations or that all exchanges must be organized as markets.

Writers often continue to be nominally independent craft workers, exchanging the product of their labors for money or promises of money in the form of royalties. Similar arguments might be made for many musicians, actors, and sports figures. In the industries of cultural production, as John Clarke and Nicholas Garnham have observed—and in areas of information production as well—corporate production and distribution typically interlock with the creative labor of "semi-autonomous or petty commodity" producers.[22] But the capitalist enterprises with whom writers, actors, and such make their contracts employ tens of thousands of wage earners as well.

"Uneven development" refers to the historical fact that the process of capitalist production did not simultaneously seize hold of all social labor everywhere.[23] Considered in both its geographic-territorial and social aspects, historical capitalism must perforce take hold progressively within particular places and spaces. Specific segments of the division of labor in households, communities, regions, and entire cultures were commodified before others. Processes of commodification also have repeatedly surrounded new means of information production: the succession of technologies of information objectification, beginning with printing and continuing through lithography, photography, film, audio and video recording, digital signal processing, and biotechnologies of genetic recombination. The suggestion that such technologies can be put to other uses must be carefully posed. In historical terms—the only terms that matter—such technologies have provided indispensable sites of capitalist accumulation. Struggles to develop alternative and oppositional uses have been undertaken more often and with greater historical importance than is typically recognized. In such instances, capital and/or states have made recurrent attempts at coercion, containment, and cooptation.[24] There have also been repeated mistakes, misjudgments, and miscues. Despite strong corporate backing, for example, the home market for videodiscs did not initially materialize as expected.[25]

Within the wider context of uneven development, the success of specific efforts to renew and extend the accumulation process is never foreordained. Successfully harnessing new technologies of objectification to extend the commodification process has typically required capital to negotiate with the social experience of target audiences or groups of users. Many cultural commodities, including, in the United States, urban daily newspapers, dime novels, films, and rock music recordings, have been carefully and productively studied with an eye to charting this complex negotiation. The social positioning of cultural commodities has repeatedly been shown to involve contingent overlaps and antagonistic refractions of shared experiences of class, gender, and race.[26]

In the late twentieth century, the first cell—production by wage labor for the market—grew to comprise an ever-increasing share of overall information and cultural production and exchange: "[T]he last fifty years have seen an acceleration in the decline of nonmarket-controlled creative work and symbolic output."[27] Commodification has become strikingly visible in sectors wherein previously it played only a limited or indirect role. For all its merits, Harry Braverman's description of the "universal market" is unhelpful in this context because it implies that the dynamic, uneven processes of market expansion are complete.[28] Capitalism has been sustained by ceaseless enlargement of markets for commodities, and this trend continues today in information and culture.

The commodification process is not restricted to cases in which information is the ultimate product or service. We must also consider information's role as a capital or producer's good. Here again, the historical importance of ever-enhanced and enlarged technical means of information production cannot be overemphasized. An increasingly intense corporate focus on "knowledge in production," often expressed in jargon like "information resource management," or "knowledge management," signifies that the prospective contributions of information to profit levels have been routinely subject to meticulous study by corporate planners and systems analysts.[29] This is true of companies producing noninformational goods and services such as automobiles and toothpaste as well as more overtly "informational" businesses. This scrutiny is widening and intensifying partly as a consequence of the enormous continuing corporate investment in information technology. Maps of the installed base of computing power in the United States in the 1980s reveal major aggregations not only in office-heavy cities like Washington, D.C., and New York but also in industrial cities such as Chicago, Detroit, and Philadelphia. Throughout this book, I emphasize that, often supported by telecommunications infrastructures, information has become an increasingly significant factor of production across all economic sectors, including agriculture and manufacturing as well as high-tech services.

The expanding and intensifying exploitation of information as a capital good expresses and adds to the larger process of commodification. The corporate search for a metric with which to measure the value and output of information production antedates the digital computer; it is nearly as old as the application of the corporate form to business enterprise.[30] Processes of corporate rationalization and work reorganization, through a steadily expanding series of applications beginning around a century ago, were based on and generated new information for management about how production occurs.[31] From a different direction, scientific and technological information grew more and more crucial to production processes and product development.[32] As described in chapter 1, these informational inputs, which in an earlier era of capitalism tended to be produced by petty proprietors, gravitated increasingly toward wage labor and market exchange. Where once they labored as independent merchants of their own labor (or sometimes as in-house employees), accountants, public-relations practitioners, bankers, lawyers, and other information-service workers became employees of giant capitalist firms.[33]

We may grasp additional attributes of the commodification process by engaging one of its most significant contemporary sites: biotechnology. Genetics and cognate fields of biology have been progressively redefined over the past half-century to accept a growing emphasis on information. "The decisive, energising perception of biology since the Second World War, the key to its strength and vigour," writes Edward Yoxen, has been one that treats organisms as information-processing machines. They begin as packets of information; they organise themselves through a process of programmed self-assembly; they operate on their environment in a controlled manner according to genetic instructions; they reproduce by condensing their structure and functional coherence into a transmittable form—that carries a message containing the instructions in a code that organisms can "read." To think of life in this vocabulary is basic to modern biology.[34]

For molecular biologists, Yoxen continues, "biology has become a kind of flatland in which the only activity is the processing and transmission of genetic information." Tersely put, "[B]iotechnology is the projection on to an industrial scale of a new view of nature as programmed matter."[35] Innovations by and around contemporary biotechnology constitute a domain of far-reaching relevance for commodification.

The transition to information capitalism does not depend on or equate with a narrow sector of media-based products. It is coextensive with a socioeconomic metamorphosis of information across a great (and still-undetermined) range. As commodity relations are imposed on previously overlooked spheres of production, new forms of genetic and biochemical information acquire an unanticipated equivalence with other, more familiar genres. Agribusinesses, pharmaceutical giants, energy and chemical corporations, and medical companies—all essentially concerned with diverse genetic and biochemical information streams—are in the midst of a continuing technological transformation of the means of information production that is every bit as relevant to our understanding as the parallel trend toward "convergence" between television, computing, and telecommunications. The connections and overlaps between genres traditionally of interest to communications research—television shows, newspaper reports, computerized data streams—and genes now subject to unprecedented manipulation and control via biotechnology compel consideration as parts of a single conceptual and historical process.

Martin Kenney made this connection explicit twenty years ago: "Biotechnology is an information-intensive technology and will very easily fit into a restructured economy based on information. Indeed, biotechnology will provide one of the new economy's crucial underpinnings."[36] Other analysts, including Jack R. Kloppenburg Jr. and Manuel Castells, have made this claim more familiar; and Bronwyn Parry has insightfully and extensively analyzed how biotechnology acts as a new information technology by virtue of "its ability to decode and reprogram . . . genetic and biochemical 'information.'"[37]

Biotechnology is the creature of a full-blown corporate capitalism. Within what Kenney calls the "university-industrial complex," the social relations of production are rapidly shifting toward market imperatives, while strictures of proprietorship mandate a continuing effort to move away from notions of science as a form of necessarily "public knowledge."[38]

This characterization, however, remains incomplete without emphasizing the profound tension between information and culture that again surfaces in biotechnology discourse. Active technical and theoretical intervention into biological information pathways resurrects an ancient usage of culture, in the sense of cultivation, "the tending of natural growth" in crops and animals.[39] Yet this usage in biotechnology is rendered profoundly problematic. What is "natural growth"? The phrase, like the passive "tending," sounds incongruous or naive in the context of aggressively invasive programs of genetic recombination and new reproductive technologies. Biotechnology thus raises the age-old boundary between nature and culture far above the threshold of habitual response.

Contending groups and interests seek control over the direction and meaning of this portentous line of demarcation, but the pace and range of this vector of information commodification are already impressive. Genetically modified food first came to U.S. supermarkets in 1994; by 2005, 75 percent of processed foods in the United States, including boxed cereals, other grain products, frozen dinners, and cooking oils, contained some genetically modified ingredients, according to the Grocery Manufacturers of America, and nearly all products using corn or soy featured genetically modified elements. The United States is home to about two-thirds of the world's 200 million acres planted with biotech crops.[40] Of course, there are many other applications in the pharmaceutical and medical industries. In and around these emergent sites of private accumulation, not only has capital sought to rework culture to further self-interested objectives, but others, opposed on varying grounds to specific biotechnology applications, have tried to make the issues amenable to political rather than purely proprietary determinations.

Biotechnology connects in other ways with more familiar domains of scholarship in culture and information, for example, by allowing us to remember that attention to agriculture and farmers informed influential early communications research into the diffusion of innovations. Kloppenburg's pathbreaking study, *First the Seed*, may center on an unfamiliar site of information commodification—plant germ-plasm—but its documentation reveals interesting and significant connections with better-known fields of communication study.[41] Intense international struggles commenced during the 1980s over the commodity status of the seed, the basic information-carrying unit of agricultural production. Specifically, UN-affiliated agencies such as the Food and Agriculture Organization became embroiled in strife over the terms of trade between developed market economies and less-developed countries in the global South. Giant transnational pharmaceutical and agricultural corporations demanded continued free access to plant germ-plasm located in gene-rich equatorial zones, while at the same time insisting that international laws of intellectual property be strengthened and harmonized to protect the profits from the hybrid seeds and drugs they sell back to these same regions. This debate carries over into debates over public health and evinces a suggestive parallel with a controversy familiar to communication researchers over the New International Information Order (NIIO).

The NIIO debate, discussed further in the next chapter, centered on the unbalanced terms of transnational production and distribution of immediately recognizable media genres: news, telefilms, sound recordings, and computer data flows. After years of effort by the United States, by the mid-1980s the movement for an NIIO was defeated. But knowledge of the development of a parallel struggle over plant germ-plasm allows us to inquire whether the NIIO defeat signals a lost war or only a lost battle. Perhaps it constitutes a preliminary phase of a more multifaceted, long-term conflict? May we not entertain the possibility that, though its ostensible subject and organizational context have shifted, the NIIO movement's underlying concerns—the systematically inequitable consequences attending capitalist exploitation of information—have simply resurfaced in what is actually a related field? I do not offer a conclusion concerning these issues; I have a different aim. The newly inclusive concept of "information," which now must be stretched to cover bioinformatics databases, and its convergence with familiar "cultural" genres such as television programs, offers a basis for analyzing unsuspected historical linkages, which may carry theoretical as well as practical political implications.

An example may be found in another apparent connection: between relentless attempts by agribusiness to turn seeds into commodities and the repeated historical success of media businesses at utilizing technologies of mechanical reproduction to enlarge the commodity sphere.[42] The crucial insight here actually lies in the contrast between these patterns of exploitation. In conventional media fields—video, musical recordings, and computer software—the attempt is to reproduce and distribute millions of copies of a standard text. In biotechnology, however, the effort revolves around supplanting a "text" with the inborn capacity to reproduce itself without human intervention—normal seeds—with one that is sterile and thus cannot do so. This difference suggests a theoretical point: mechanical reproduction per se should not be employed as an analytical fulcrum because it is reproduction in the service of capital accumulation that is crucial.

However portentous, posing the issue of commodification solely in contemporary terms is to accept, at least implicitly, the postindustrialists' claim that the social discontinuities in which we are ostensibly enveloped can lay claim only to a grotesquely brief history, beginning with the postwar rise of digital microelectronics. The technological determinism underlying this formulation is insupportable and comprises a point of considerable vulnerability. Postindustrialism's proponents and sympathizers have tacitly acknowledged as much by making repeated efforts to tone down or mitigate the theory's overriding emphasis on the suddenness of the supposed contemporary socioeconomic break. Lacking such a historical foundation, Bell again provides a seemingly authoritative, but actually evasive, disclaimer: "The postindustrial society is not a projection or extrapolation of existing trends in Western society; it is a new principle of social-technical organization and ways of life."[43]

Despite these inadequacies, no satisfactory historical formulation has been reached. Even scholars who work with ideas of information commodification have tended to limit their studies almost exclusively to contemporary (postwar) society. Understanding what is at stake in trying to specify the historical character of the process therefore is difficult. If information commodification can be shown to have an identifiable history, then the massive discontinuity invoked by postindustrial theory collapses. More important, it becomes possible to use knowledge of how, often in hitherto undetected ways, information commodification has contributed to and become interconnected with continuing processes of capitalist development. To work toward this more encompassing historical revision, we need to ask why, how, where, and when have wage labor and the market taken hold within information and cultural production as well as in agriculture and manufacturing? If the goal is to integrate information fully into the history and theory of capitalist development, moreover, then what sort of periodization scheme will allow us to shed the most light on the multifaceted and uneven process of commodification? The remainder of this chapter shows that attempts to grapple with this hidden history are likely to engender substantial revisions to received historiography.

The Longer Revolution: Historical Origins of Information Commodification

Received scholarship locates the origins of information and cultural commodification in eighteenth-century England. This chronology emanates from long-standing research traditions, repeatedly reaffirmed most influentially, and proximately, in studies of authorship and copyright law. This dating will not prove satisfactory. But this scholarship is of arresting interest in its own right and opens vital entry points for consideration of basic conceptual and periodization issues.

Studies focusing on the emergence of European copyright law and linking its rise to changing conceptions of authorship often cite Michel Foucault as an inspiration.[44] They were no doubt also influenced by the rapidly increasing visibility of "intellectual property" and the policy conflicts around it.[45] In any case, the historical bond between author and work became freshly problematized. Jane Gaines summarizes this research by observing that originality in the work—the basic prerequisite for extending legal protection—began to be "guaranteed by the notion of the individual as subject"; copyright was based on the author's ostensible "singularity, his separateness from other humans."[46] This shift had been dated, by Martha Woodmansee, to the mid eighteenth century,[47] when ascending romantic notions of authorship as poetic genius began to offer what the legal scholar James Boyle terms "a conceptual basis and a moral justification for intellectual property."[48] Mark Rose details how copyright law could then bind together the newfound subjectivity of the author with the emergent objectivity of the literary work.[49]

This shift in the literary system harbored far-reaching implications. Subsequent practices of cultural production, often overtly collective rather than individual, nonetheless were channeled repeatedly by their proprietors, especially when they could rely on accepted systems of notation, into this legal vehicle for the protection of individual private-property rights. Thus, as Boyle emphasizes, "[T]he romantic vision of authorship continues to influence public debate on issues of information—far beyond the traditional ambit of intellectual property."[50] Jeanne T. Allen argues that at least some cultural practices that were unable to lay claim to strong copyright protection met with adverse consequences in the marketplace.[51]

In truth, however, the significance of this newfound chronological link between copyright and altered conceptions of authorship, goes far beyond this. Originating in and around the book trade, the legal right to private property in information emerged first in England, where modern copyright statute dates to 1710.[52] Mark Rose has adeptly shown that issues raised but left unresolved by the 1710 act triggered a protracted literary property debate, which reached a definitive culmination only in 1774.[53] This periodization possesses a much larger, if mostly implicit, importance: it appears to signify that author and work came to interlock in an enduring system of institutions, legal forms, and literary practices—the substrate of modern culture industry—at a formative historical moment, the moment of industrial capitalism's rise in England.[54]

This argument, which had already long since been made explicit, exhibits considerable virtues. The fiction of individual authorship helped enable the culture industry to grow beyond its modest beginnings on Grub Street. Not books alone but, through an uneven progression that has consistently enlarged their scope and significance, drawings, theatrical performances, photographs and films, sound recordings, videos, and computer software, to mention only some of the most prominent forms, came to bear the legal impress of copyright. Private ownership rights were thereby fixed in particular cultural commodities, not only helping capital to exact monopoly rents from consumers but also to siphon surpluses from the wage earners and independent contractors whose collective labors produced these commodities. Analogous historical arguments for patents,[55] trademarks, and trade secrets have not yet been merged in a full-scale economic history of intellectual property, but work in exactly this direction is beginning to appear.[56]

The growing prominence of copyright over the last two centuries is incontestable. This alone provides a salutary reminder that information and cultural commodification extends farther into the past than the postindustrialist account allows. However, does the law of copyright, or even of intellectual property in general, actually comprise the decisive, let alone the exclusive, basis for commodification? Is today's sprawling culture and information industry essentially the result of a newly conducive legal or authorial regime? Does a focus on the law and the author provide sufficient analytical room for capital—for the publishers, film studios/distributors, television networks, software firms, and biotechnology companies that control the means of cultural and informational production? What about the workers, whose skills are engaged by these capitalists to create this unfolding range of goods and services? To what remote corner does a formal fixation on law consign the relations of cultural and informational production?[57]

One reason for questioning attempts to locate the origin of decisive change in eighteenth-century English law is that capital accumulation in the book trade had already ripened for over two centuries previous to passage of England's copyright law, generating for individual publishers estates rivaling those of nearly any field of capitalist endeavor.[58] The law is better seen as one moment in a still unchronicled longer history, than as the essential vector, or primary site, of transformation.

I am not asserting that law is merely an epiphenomenal superstructure. "Intellectual property" law possesses a crucial function; absent these state sanctions, it would be far more difficult for capital to restrict the uses and users of information and to expropriate surpluses from at least some groups of primary producers. As Allen argues, "[T]he legal structure's support for certain entertainment commodities" over others has repeatedly proven significant for their success or failure.[59] Explaining the victory of the early U.S. film business over its rivals in vaudeville and live theater, she poses the issues this way:

> In the competition for business success, mass culture industries sought and responded to legal decisions which determined what aspects of their property could be controlled and defended and hence what pursuits might reward the considerable financial investment required. Did one form of mass entertainment offer greater control? greater opportunity for monopoly protection?[60]

I step back from further consideration of these questions to return to periodization. A focus on relations of production in culture and information rather than only on law allows us to ask, Should eighteenth-century England be construed as the historical launchpad of the commodification process with respect to culture and information?

Early and enduringly influential postwar studies explicitly lodged the rise of characteristically modern genres in eighteenth-century England;[61] they also identified modern-sounding debates over the forms and meanings of these concurrently changing cultural practices.[62] In Raymond Williams's phrase, a "long revolution" unfolding in culture and communication—education, drama, the press, and language—was related to transformations commencing simultaneously in the English economy and polity: the Industrial Revolution and the bourgeois democratic revolution.[63] The practices and forms of culture thus comprised for him a third profoundly formative and intertwined but neglected dimension of secular change.

Stuart Hall later furnished this seemingly comprehensive sketch in an attempt to specify a historical basis for the characteristic forms of bourgeois cultural hegemony:

> The modern forms of the media first appear decisively, though on a comparatively minor scale as compared with their present density—in the eighteenth century, with and alongside the transformation of England into an agrarian

capitalist society. Here, for the first time, the artistic product becomes a commodity; artistic and literary work achieves its full realization as an exchange value in the literary market; and the institutions of a culture rooted in market relationships begin to appear—books, newspapers, and periodicals; booksellers and circulating libraries; reviews and reviewing; journalists and hacks; bestsellers and potboilers. The first new "medium"—the novel, intimately connected with the rise of the emergent bourgeois classes (cf. Watt, 1957)—appears in this period. This transformation of the relations of culture and the means of cultural production and consumption also provokes the first major rupture in the problematic of "culture," the first appearance of the modern "cultural debate" (cf. Lowenthal, 1961).[64]

Raymond Williams was equally certain of the eighteenth-century chronology in a discussion of the drama: "The eighteenth-century middle class broke up the old forms, which rested on meanings and interests that had decayed. Alternative forms were created, but were relatively isolated or temporary in their success. Only much later, when the class had built its major social institutions, was there an effective turning towards the making of a distinctive cultural tradition, at all levels of seriousness."[65] This received view was powerful in part because it was positioned as a necessary supplement to an already well-embedded historical conception. Belief was widespread that a "dual revolution" erupting in the late eighteenth-century economy and polity rendered Europe, as represented first and foremost by England, a prototype or paradigm of the modern process of development.[66] Industrialization and democracy, in this view, comprised the established twin pillars of modern society. The cultural realm, in revisions offered by critics such as Williams, constituted a third—necessary, but in this sense supplementary. A sustained emphasis on the eighteenth century seemed only to strengthen this formulation.

If we rely on this periodization, capitalism itself could appear, with seeming rectitude, mainly through the ubiquitous and well-established figure of the ascending middle class, the hegemonic bourgeoisie, or, as for Ian Watt, the growing primacy of economic individualism. Additionally sensitized by their sometime awareness of neighboring France, critics endeavored with greater or lesser explicitness to frame English cultural practice within the terms of a presumed governing conflict between this emergent social class and a sharply differentiated, declining aristocracy.

There can be no question that this ensconced critical tradition generated potent insights. And it has been extended to a broadening array of eighteenth-century cultural practices, including sports, commercial amusements, and consumer goods.[67] Unquestionably, eighteenth-century England witnessed a significant reorganization and growth of capitalist cultural production. Yet the analytical and historical roots of this efflorescence are insecure. Its chronological placement requires reexamination, as does its portrayal of the class dynamics of historical change around information and culture and more generally.

The metamorphosis of England into an "agrarian capitalist" society—to which Hall rightly refers in attempting to contextualize the emergence of culture industry—began not (as he states) during the eighteenth century but, according to an influential body of historical scholarship, much earlier.[68] It was from the fifteenth to the eighteenth centuries that the crucial triadic relationship between landless wage laborer, capitalist tenant, and commercial landlord powered its way into English agriculture. During the eighteenth century the long transition was completed, as agrarian capitalism attained a triumphal maturity.[69] A thoroughly capitalist foundation actually preceded subsequent capitalist industrialization.[70] Presiding over the development of English agrarian capitalism, therefore, was not an emerging metropolitan bourgeoisie but a fluid social amalgam of improving landlords and proprietors, whose membership drew on aristocratic as well as bourgeois elements. The pivotal social relationship was not the storied contest between aristocracy and bourgeoisie but that between agrarian capitalist landlords and improvers, on one side, and exploited agricultural wage labor, on the other side.

This historiographic shift has yet to be integrated into the received account of English cultural development.[71] The received account's effort, often laudable, was again to trace the historical sources of hegemony: how culture industries built outward in their effects toward other social groups, from their initial foundation within the subjectivity of the eighteenth-century metropolitan bourgeoisie.[72] The reality of capitalist hegemony is not at issue; but its historical basis, social origins, and developmental dynamic must be opened to question. Are we not foreclosed from any simple equation between commodification across its range and the class interests, cultural preferences, or consciousness of an already formed metropolitan middle class? New emphasis must be placed and fresh evidence gathered on the neglected social relations of production, through which capital's own prior development occurred.

Adherents of the standard view proved no more attentive to the specific conditions of cultural practice than to the overall social relations of production they helped constitute. They typically emphasized form and genre. Hall, for example, is quick to generalize to the artistic product, but his examples of a culture rooted in market relationships connect the commodification process exclusively with literary work and the literary market, and even with specific genres, the novel and journalism. Both these genres, far from coincidentally, tended at this stage to be associated by scholars overwhelmingly with the middle class. All this resulted in serious errors of omission.

Vast reservoirs of culture and information are tacitly marginalized through a reliance on conventional literary notions of genre: cadastral maps, scientific illustrations, popular magic, and government records. Overtly collective practices of performance likewise tend to be neglected. Had Hall chosen, for example, to analyze drama—the subject of an especially nuanced chapter by Williams[73]—his depiction of the origins of modern media would have required significant revision. By the Elizabethan era (the late sixteenth century), with the arrival of permanent performance spaces (theaters), Laura Taub shows that capitalist commodity relations began to penetrate and transform the drama.[74] If we step back from the novel and foreground publishing as a whole, we find that it too was organized on decisively expansionary capitalist lines from its inception in late fifteenth-century England and elsewhere.[75] Important cultural forms were not simply in the wings, awaiting the eighteenth-century triumph of the urban middle class; centuries earlier, they had already begun to forge new circuits for capitalist commodity production. In the case of drama, as Taub relates, they may even be considered spearheads of capitalist development.[76] Through this neglect, therefore, the received view truncates and miscasts the history of culture and information and of capitalism itself.

Williams came to acknowledge that he had previously underestimated the role of news organs in the revolutionary era of the seventeenth century. In 1965, he shifted his periodization somewhat, but still hedged, presumably with a view to the weight of accepted historical evidence: "The decisive social origin belongs there, but, as with so many other initiatives of that time, development was thwarted by the whole political and economic situation, and something like a new beginning had later to be made."[77] Later research raises additional questions about whether such genres as the newspaper or the novel—long understood as congealed expressions of instrumental need and class subjectivity among a self-determining bourgeoisie—may themselves have sprung from earlier and more multifaceted and conflicted origins.[78]

Cultural and informational commodification commenced not after but within the acute social struggles marking the transition to capitalism. It transpired not as a benchmark of an already secure mode of production, as the eighteenth-century periodization implies, but within the context of turbulent and often ambiguous and overlapping social class relations, accompanying the earlier ascent of capitalist commodity production within English agriculture. The class origins and affiliations and the consequences of commodified cultural and informational practices were almost certainly more complex and more contested than the customary account suggests.[79]

The established association of commodified culture and information production with capitalism's emergence has been placed in question. If we accept the received view, we must try to explain why the culture industry lagged behind the emergence of capitalism by hundreds of years. If we reject the standard view, other evidence must be adduced that will link commodification of cultural production more fully to that long moment of transition and the social relations that circumscribed and animated it. Commodification must be situated in terms of the dominant relationship of exploitation within contemporary English society: that between agrarian capitalists and landless wage laborers. We should expect this paramount social relationship to be expressed, refracted, and referenced in multiple ways and at a range of sites. The work of making these connections, however, largely remains to be done.

I have criticized postindustrial theory for substituting a misconceived focus on the transcendence of industrial society for what in fact has been a historically continuing process of commodification. Now I will add that the received stress on industrialism is doubly flawed because it also misconstrues the origins of the phenomenon it purports to study. The commodification of culture and information has been a continuing, if uneven and conflicted, process throughout the duration of capitalist development. With this insight, it should be possible to glimpse in bold relief how the commodification of culture and information accelerated during our own historical epoch.

3

Accelerated Commodification

Though perhaps not the first such historical moment, a protracted episode of accelerated information commodification commenced around 1970. We have long known that the key impetus to this rapid enlargement of wage and market relations came from a systemic crisis.[1] Throughout the United States and, to a varying extent, the rebuilt economies of Europe and Japan, surplus productive capacity and resurgent competition overtook a series of often unionized manufacturing industries, from steel and automobiles to TV sets, chemicals, and textiles.[2] The U.S. share of global steel production declined from nearly 50 percent in 1950 to 20 percent in 1970, part of a continuing relative decline in the U.S. share of global manufacturing output.[3] A profit slowdown and creeping stagnation were the portentous results.

Recognizing a need to develop new profit sites, corporate executives and politicians mobilized to craft a response. This was not the first time that overproduction and competition engendered efforts by elites to rejuvenate the market system, but the pivotal role accorded to information and communications as a solution was unprecedented.

As a result of its hothouse growth throughout the early postwar era, by around 1970, the dynamism and the mounting economic and strategic importance of this sector were increasingly acknowledged. Deliberating in 1970–71, the Senior Executives Council of the Conference Board—a leading U.S. business policy organization—concluded, "The information 'industry' in its broadest sense could soon become the leading edge of many economies. . . . It is now a matter of highest priority that we begin to perceive and conceptualize information technologies, industries, and resources in comprehensive, or what might be called strategic, terms."[4] A congressional committee in 1980 reiterated what had become common wisdom, declaring that, in the context of an unabating economic slump, "computers, telecommunications, and the services which grow out of or depend on those technologies . . . are the critical industries for continued economic growth."[5]

What special features did communications and information possess, however, that qualified them for the task of lifting the market system out of its lethargy? If the need was to break the slide into stagnation by identifying fresh fields of market development, then the question becomes: How could information and communications be reorganized to establish a new wellspring for the accumulation process by reviving profitable corporate growth?

Undercutting Demands for a New International Information Order

As these questions were fully engaged, elites concurrently had to grapple with a related problem: an international political challenge to the existing, already skewed, arrangements for global information provision.

Led by India and Indonesia, more than a score of newly independent nations in Asia and Africa joined together at Bandung, Indonesia, in 1955 to promote "world peace and cooperation," to consider "problems affecting national sovereignty and of racialism and colonialism," and to develop forms of cooperation in their common interests.[6] Subsequently joined by dozens of other poor nations, including many Latin American countries, this "Third World" bloc began to press for basic changes in the structure of the international political economy. During the 1970s, it fed into the Non-Aligned Movement (NAM)—so called because members tried to create space for national development between the United States, on the one side, and the Soviet Union, on the other side—and the NAM's reform agenda assumed mounting visibility and importance. In this wider context, the NAM (often supported by the Soviet Union) fastened increasingly on communications, culture, and information as a vital problem. A New International Information Order (NIIO) was needed, NAM countries began to insist in the United Nations Educational, Scientific, and Cultural Organization (UNESCO) and other forums, to help remedy dominative structures and unequal relationships. Absent fundamental changes in the system of global information provision, the chances of attaining the NAM's preeminent goal—a New International Economic Order—were lessened dramatically.

An array of inequities was ultimately identified: mass distribution for global audiences of news full of false or distorted images of poor nations and peoples; the virtual monopoly over the electromagnetic spectrum exercised by the developed market economies; the imposition of copyright charges on publications entering Third World countries; the lack of anything approaching inclusive communications infrastructures and services; the undermining of indigenous audiovisual production through the dumping of U.S. television programs and films; systematic violation of sovereignty via emerging supranational communications systems, principally satellites, and networked transborder data flows controlled by foreign states and supranational corporations; and, decisively, cultivation through the introduction of corporate-commercial media systems of consumerism in place of development priorities that would improve access to education, food, and medical care.[7] Without profound changes in international communications and information to remedy these faults, NAM countries insisted that no path toward genuine political independence and economic development could be cut.

This confrontation sharpened during the first half of the 1970s.[8] To simplify a complex history, a key point in the Non-Aligned countries' platform—articulated by Mustapha Masmoudi, the information minister of Tunisia—was that "[i]nformation must be understood as a social good and cultural product, and not as a material commodity."[9] The U.S. policy stance was directly contradictory: "In international commerce, information and communication are commodities, whose value is rising as the recognition of their importance grows."[10]

Were the demand for an NIIO to prevail, the result would be to divert and potentially to block attempts by the United States, then unquestionably the world's paramount economic, military, political, and cultural power, to recast information into a general foundation for profit-seeking market expansion. By around 1980, as they became less preoccupied with Vietnam and Watergate and increasingly cognizant of the need to combat stagnation, U.S. elites moved from defensiveness to confrontation. President Reagan, sychronizing this offensive with Britain, pulled the United States out of UNESCO at the end of 1984. Debt crises in many countries and structural adjustment programs imposed by the International Monetary Fund on dozens of poor nations throughout the 1980s and 1990s granted new leverage to U.S. leaders. Dramatic geopolitical changes bequeathed windfall opportunities to exert further pressure and to extract additional concessions. The abrupt fall of Soviet Socialism and the more gradual reintegration in the transnational market system of Communist China and nationalist India stripped away critical support for the NAM countries' attempt to pursue alternative developmental directions. In short, the NIIO movement was defeated, and the shifting framework of global power relations became conducive to a sustained counterthrust by the leaders of the developed market economies (DMEs)—at the center of which was a drive to broaden and accelerate information commodification.

Before turning to inspect leading features of this multipronged strategy, two basic points merit emphasis. First, policy moves to cultivate information's profit potential in hardware, software, and services were opportunistic, not executed according to some preconceived blueprint or long-range master plan. Second, however, the corporate focus on information was broad-based from the beginning, encompassing more than the industry sector that was burgeoning around the supply of communications and information. The power of the corporate state's push into information and its attempts to use information as a new basis for profitable growth expressed a *general* imperative. Information organization, processing, storage, packaging, and distribution comprised ubiquitous corporate functions, and companies engaged in information production and use spanned *every* economic sector, declining and emergent, from agriculture and mining and manufacturing to services.[11]

Government's Foundational Role

In the supply of innovative information and communication technologies (ICTs), as opinion leaders in other developed countries perceived with alarm,[12] throughout the early postwar decades U.S. big business built up a commanding comparative advantage. Continuing torrents of government research and development funding for military contractors underpinned virtually all U.S. corporate ICT development.[13] Technological innovations by the military-industrial complex during and after World War II were incorporated into a succession of systems for capturing, storing, sorting, sharing, and *measuring* information, increasingly in real time.[14] Computer hardware and software, satellites, packet-switching, and online information-retrieval services (such as DIALOG, Lexis-Nexis, Orbit, and BRS [Bibliographic Retrieval Services]) all originated largely in military and NASA space contracts.[15]

As the class power of an ascending fraction of capital increased, the corporate chieftains of informationalized capitalism pressed new demands on the state, and the government responded by documenting, elevating, and projecting these policy preferences domestically and internationally. Widely hailed as a return to the supposed natural logic of the free market, therefore, accelerated commodification was anchored for strategic purposes by the state. Government intervention was pivotal and sustained: not only in procuring continued research and development funding but also for telecommunications industry liberalization; privatization of what had been public information; strengthening legal rights to private property in information; and shifting global trade and investment rules to favor services. Each of these initiatives contributed importantly to a sweeping process of accelerated commodification.

A New Capital-Logic for a Modernized Information Infrastructure

Let us look first at network infrastructures, which had to be continually built out, upgraded, and modernized for informationalized capitalism to progress.[16] On what institutional basis did this metamorphosis proceed?

Changes in and around computer networks unfolded as part of a multifaceted transformation of the ground rules that structure telecommunications and electronic media systems. Generating massively enlarged informational inputs and outputs alongside an enlarged need for coordinating complex business processes across dozens of locations, big corporations' active integration of ICTs to support ever extending production chains was the predicate of supply-side market growth by vendors like IBM, MCI, and Cisco.[17] Corporate expenditures on network hardware, software, and personnel correspondingly rocketed,[18] from Citicorp's ATM networks to Wal-Mart's inventory database[19] to Nestlé's toll-free hotlines, which were installed in 1992 to compete with rival Procter and Gamble and by 2003 generated 880,000 calls a year.[20] As ICTs grew to become the largest category of corporate capital investment—depending on how they are tallied, between 35 and 60 percent of the U.S. total by the early 2000s—surging corporate ICT outlays began to drive overall economic growth.[21]

Cascading corporate ICT investment also engendered an incremental but ultimately radical overhaul of the structure and policy of telecommunications. In chapter 5, I scrutinize the contradictory nature of this new trend; here, I simply survey its aim and general contours. Beginning in the 1970s, U.S. policy makers targeted the arrangements for controlling global telecommunications infrastructures. "Services-related industries are information intensive," noted a Reagan-era U.S. report, "and thus depend heavily on advanced communications and computer systems to provide necessary access to and transfer of information. The strong tie between the information and telecommunications sectors extends beyond the domestic sphere . . . to international markets as well."[22] To open a vast new field to capital, to advance and enlarge markets for new information technology hardware, software, and services, and to develop and extend networked business processes, the United States worked to transform global telecommunications.

Elites worldwide increasingly shared these goals. Policy makers throughout the developed market economies of Western Europe and Japan—which had long organized and operated their telecommunications systems as government ministries of posts, telephones, and telegraphs—were cajoled, and some arms were twisted; but they proved ready and willing to embrace privatization and liberalized market entry as prerequisites of economic "competitiveness." Throughout the global South, more aggressive tactics were employed. After decades of indifference, DME policy institutions now elevated to the status of orthodoxy the tenet that national economic development was being stifled by inadequate telecommunications. The World Bank, the International Monetary Fund, the World Trade Organization, alongside the U.S. Trade Representative, the Federal Communications Commission, and other apostles of neoliberalism, used multiple levers to pry open global networking to corporate-commercial investment.

During the 1980s and especially the 1990s, the largest and fastest sectoral reorganization of productive assets in world business history took place, loosely synchronized to make-over what had been a (typically inadequate) public service into a corporate-commercial function. Privatizations by the dozen were undertaken, with the U.S. service sector—banks, law firms, accounting companies, and public relations agents—capturing a disproportionate fraction of the fees dispensed for managing the process.[23] Immense sums of capital were mobilized and directed toward this fresh investment outlet. Sectoral stagnation gave way to astonishingly rapid network build-outs around advanced digital technologies, including, by the mid-1990s, the Internet. By 1997–99, fully half of global telecom investment was being absorbed by "developing and transition" countries—that is, countries not included in the Organization for Economic Cooperation and Development (OECD).[24] By 2002, 49 percent of the world's wireless and wireline connections were located in the less-developed countries (still a disproportionately low figure, because they possess nearly three-quarters of world population).[25] Investments in capacious transoceanic systems were a vital complement, as they permitted transnational corporations to reach deeper into newly networked national economies in search of markets, natural resources, and cheap labor.

Contemporary "globalization" stemmed in no small part from the innovation of networked business functions, which underlay lengthening transnational supply chains that took profit-maximizing advantage of low-wage labor pools. The continuing relocation of manufacturing to the global South, especially to China, was the earth-shaking result; between 1980 and 2000, as corporations developed networked global supply chains, their direct investments in plants located in less-developed countries helped push the share of world manufacturing value added in less-developed countries from 14 to 24 percent.[26] Enhanced transnational corporate profitability was the object— and, of late, the result. U.S. transnationals, for example, took in $315 billion of profits from overseas operations in 2004, up 26 percent from 2003.[27] The movement of manufacturing to the global South has showed little sign of slowing; in 2005 transnational corporations continued to look to China, India, Latin America, Eastern Europe, and the Middle East to build markets, and General Electric expected as much as 60 percent of its revenue growth over the decade beginning in 2005 to come from less-developed countries— up from 20 percent in the previous decade.[28] Foreign direct investment into selected poor countries surged.[29]

The gale of creative destruction that blew through telecommunications comprised only one aspect of a wider metamorphosis. U.S. policy makers, according to a Reagan-era report, would "strongly oppose any actions that would interfere with the ability of producers and users to make optimum use of information as a productive resource."[30] In this formulation, "producers and users" means corporate—above all, transnational corporate—producers and users of information. Turning next to the supply of information software and services, let us revisit the question: How was it that informational activities offered a singular potential for profit making, and how has this potential been realized?

Expropriation of Nonproprietary Information

The major route to profitability was—as the case of telecommunications shows—by annexing existing areas of social investment, clawing them into the corporate-commercial domain, and, concurrently, transforming their character. This historical movement was nothing less than a sweeping and multifaceted process of expropriation.

The huge scale on which the world's information resources already were being cultivated is typically forgotten. Institutions for information provision, however, had been widely formalized, built up at collective expense, and put in motion by skilled social labor. Schools and colleges, government agencies, post offices, museums, and libraries collectively constituted one axis of this system. Family and local community-based stocks of indigenous, vernacular, or traditional knowledge about farming, healing, and learning constituted a second axis throughout the global South. These information programs and traditions, taken together, existed as an enormous collective reserve of not-for-profit activities, and they played indispensable roles within their home cultures.[31]

If parts of this assemblage could be reorganized along proprietary lines, then corporate capital could expand into a largely untapped but often highly developed realm of sociocultural labor. Elite programs of political-economic reconstruction thus turned increasingly around the objective of enclosing the immensity of global communications and information provision, a project that—though it remains unfinished—should be seen as a species of what David Harvey calls "accumulation by dispossession," the paradigm for which was set via enclosure of common lands in England during the epochal transition to agrarian capitalism hundreds of years ago.[32]

Accumulation by dispossession for informationalized capitalism dates back at least to the First World War, as Michael Perelman points out.[33] During the war, several thousand industrial patents held by German firms—notably, in the chemical industry, which German capital then dominated—were expropriated by the U.S. state and turned over to burgeoning U.S. rivals (via the American Chemical Society).[34] The formation of the Radio Corporation of America as a high-tech "chosen instrument" of U.S. state policy, similarly, was based on the takeover of German patents and British radio-system assets in the United States.[35] The Second World War witnessed the development of a more systematic approach, relying upon what one participant called "vacuum-cleaner methods." Beginning immediately after Germany's defeat in 1945, as authorized by executive orders by President Truman, vast quantities of proprietary German industrial, technical, and scientific information were expropriated—and hundreds of German scientists brought to the United States—by U.S. occupational authorities working hand in glove with representatives of large, high-tech U.S. companies.[36] The overall value of this booty has been estimated by the historian John Gimbel at around ten billion in 1945 dollars—an immense sum.[37]

After around 1970, original accumulation for informationalized capitalism was pursued in the increasingly encompassing context of neoliberalism: through systematic preferment of corporate commercial interests, especially over earlier state programs for social welfare. Funding and service cutbacks were mandated for public-sector institutions such as schools and libraries, while privatization of not-for-profit service providers and widespread outsourcing bolstered corporate capital. Year after year, government information programs were raided and shut down by would-be private suppliers.[38] The U.S. Postal Service, which attempted to develop a form of electronic mail and to offer it within a more-encompassing system of public provision, was preempted as corporate carriers targeted emerging markets for electronic communications.[39] Schools and universities were caught up in a complex and unfinished commodification process, which undercut the mechanisms for providing this public good; symptomatically, by the early 2000s, more students were enrolled at the University of Phoenix, a for-profit education provider, than at any other U.S. institution.[40] Museums and archives acquiesced to mushrooming corporate sponsorships and direct marketing initiatives—or sometimes simply auctioned off pieces of their collections to private collectors, such as the Wal-Mart heiress who was the high bidder for an Asher B. Durand painting sold by the New York Public Library.[41] The U.S. Government Printing Office's long-standing and hard-won role in producing and distributing government documents, already eroded by prior campaigns against it, came under renewed attack. The executive branch wanted the GPO to adopt a policy of permitting individual departments and agencies to let contracts to commercial bidders—a policy that would return the nation to the crazy-quilt of often corrupt practices that prevailed before the U.S. Civil War, to which the GPO was the historical solution![42] Libraries experienced repeated incursions by for-profit rivals, who try to take advantage of them as a captive market.[43] Scientific and technical journals were acquired from professional societies or freshly established by information conglomerates such as Reed-Elsevier, companies that then sabotaged the practice of public science by imposing exhorbitant charges on libraries and other subscribers. Efforts to confer windfall profits on private satellite companies by doling out government contracts for the space imagery they generated commenced during the 1970s and continued intermittently thereafter.[44]

These initiatives represent a consistent attempt to discredit, to attack as illegitimate, the very principle of nonproprietary information provision. More than twenty-five years ago, the director of what was then called the Information Industry Association made the startling—but deeply revealing—assertion that libraries had imposed "an iron curtain of free information" over the land.[45] Albeit often inadequately, the existing system still demonstrated the feasibility of attempting to supply society's information needs as collective goods offered on public-service lines. For commodification to proceed, the principle of social provision had to be supplanted—albeit, in practice, only where profitable—by the idea that commodity exchange should dominate in the construction of new consumer and business information markets.

When a government information function or service was charged with "competing with the private sector," this was characteristically because corporate-commercial suppliers had newly decided to target the function for profit making. The decision to privatize what had been built up and paid for as public information might be made by an affiliate of one of the trade associations that serve the corporate information business—the Software and Information Industry of America, the Business Software Alliance—or simply by an individual corporation. A proceeding was then undertaken to determine whether unfair competition actually existed, in which corporate representatives had a strong say while the larger public was kept almost entirely in the dark; in some cases, public opposition to a corporate takeover was simply ignored. Once dissent was preempted or finessed, the government service was folded, sometimes by handing it over to its newfound corporate rival.

In these diverse contexts, not only products or services or even whole organizations but social labor was commodified. As capital took over and redirected this previously uncommodified social labor, a further effect was to weaken and destabilize the public sphere. As Frank Webster specifies, not principles of social well-being or democratic rights but "market criteria, i.e., whether something makes a profit or loss, are the primary factors in deciding what, if anything, is made available."[46] This was not simply a shift in organizational practice, therefore, but an expropriation animated by a basic transformation of priorities. This entire process was then bolstered and ratified through an equally radical transformation of law.

Extending and Enlarging Private-Property Rights in Information

Laws of copyright, patent, and trademark were presented by state authorities as entirely benevolent. In actuality, however, their function is far from innocuous. As Michael Perelman and others have argued, it is to permit business to assert legal forms of monopoly power.[47] Seventy years ago, this function was viewed with considerable mistrust and antagonism; patents still were widely viewed as a privilege granted to inventors by the commonwealth with the expectation that this limited monopoly grant will generate reciprocal improvements for society as a whole. As an official government report emphasized in 1941, "'[T]he patent derived its value from the contribution which the invention made to the general welfare . . . privilege was the instrument of policy.'"[48] Because unworked and opportunistically licensed corporate patents were widely seen as having contributed to the Great Depression, reformers sought to break the hold exercised by major corporations over them. At least some restraints were imposed. It is profoundly ironic that the rise of informationalized capitalism was predicated on the New Deal welfare state's nonproprietary patent-licensing policies, which among other things ensured that the transistor, a key building block of microelectronics and ultimately of the Internet, was free for all to use.

In contrast, during the 1980s and 1990s corporations advanced increasingly systematic claims to what they now succeeded in relabeling "intellectual-property rights." These "rights" were then strengthened and generalized spatially, to cover the earth, and socioculturally, to confer additional corporate control over an expanding array of products and labor processes.

Again, political intervention by states was required. A U.S. trade diplomat conceded that, as late as the 1980s, "few of the world's developing countries even had intellectual property laws."[49] To address this perceived deficit, the U.S. government and corporate leaders mobilized, to shift an arcane legal field to the forefront of global economic policy making.[50] The world then witnessed "the greatest expansion ever in the international scope of intellectual property rights," as, during the 1990s, "dozens of countries strengthened their intellectual property laws and regulations (often under pressure from the United States)."[51] Susan K. Sell relates that the global regime that was established to legally constitute private property in information (mostly via the General Agreement on Tariffs and Trade and its successor, the World Trade Organization) was not the result of mutually sought multilateral agreement but of activism originating with a mere handful of determined executives representing pharmaceutical, entertainment, and software companies.[52] But the range of corporate beneficiaries is encompassing.

Intellectual-property laws stake out and seek to enforce artificial information scarcities; they constitute not defensive but preemptive claims. Copyrights afford corporate owners monopoly power for a span of many decades over an ever increasing list of cultural properties; a onetime president of Viacom, a top multimedia conglomerate possessing thousands of copyrights, boasted that "'Viacom is fundamentally a software and copyright-driven company.'"[53] Trademarks brand consumer goods and services with proprietary insignias, hoping to play on consumers' brand loyalties to generate outsized corporate profits especially coveted during periods of stagnation; Time Warner has registered two thousand trademarks connected to licensed merchandising around the Harry Potter brand.[54] Corporate trademark owners achieved an especially noteworthy success when, at the expense of non-proprietary interests and concerns, they were able to remake the Internet domain-name process into a trademark owner–friendly function.[55] Patents grant to their corporate holders a third species of monopoly power, deployed far more extensively—and unevenly—than would have been true had they been used only for the "useful" inventions of a disinterested science.[56] Corporations producing everything from seeds to television shows, from sneakers to networking software, have joined a pan-corporate scramble to patent, copyright, or trademark anything in sight.

In 2001, the U.S. issued around 160,000 patents; Brazil issued around 3,500.[57] By comparison, in 2001, IBM (the U.S. patent leader) obtained 3,411 U.S. patents—nearly ten times as many as were issued to people in Arab countries during the two decade interval 1980–2000 (a total of 370 industrial patents).[58] The United States and other rich industrialized nations hold 97 percent of worldwide patents.[59] Some 80 percent of the world's research and development and a similar share of its scientific publications come from the developed nations.[60] The newly widened forms of privilege and power that underlay these disparities have produced what economists call "comparative advantage," but this comparative advantage must be seen as a systematic political-economic achievement.

Once more, this constitutes active expropriation: Efforts to cement private-property rights in information are essential to realizing profit from the new commodities. Editorially attacking the European Commission's ongoing investigation of Microsoft's market power (in the aftermath of the Bush administration's sellout settlement of an antitrust case against the software giant), the *Wall Street Journal* supplied a sense of the hard edge of U.S. policy on this vital issue: "Somehow the U.S. government has to draw a line in the sand here. Intellectual property and services are the growth engine of the modern economy, of which Microsoft is a crown jewel."[61] Strengthened intellectual-property rights continue to function as a prime objective of big business as a whole. As long as they are seen as contributing to accelerated commodification, however, policy actions tend to be framed as if they are occurring in some neutral technocratic space, beyond the play of corporate and state power. As I write, for example, news reports observe that India, often an opposing voice in debates over private property in information, has altered its national patent laws to allow greater proprietary control by corporations as a condition of entry into the World Trade Organization.[62]

The profits for legally constituted private property in information are high. Sales of drugs developed from natural compounds sourced mostly in less-developed countries, for example, recoup to pharmaceutical companies a murderous $75–150 billion annually.[63] Over half a century, between 1950 and 2002, the profitability of the U.S. brand-name pharmaceutical industry has tended to exceed the average for all industries by a large margin; between 1998 and 2002—admittedly, a high point—its average net earnings as a percentage of sales was about 16.8 percent, compared to 5.6 percent for all industries.[64] However, via inflated prices, patents, copyrights, and trademarks extract a variable and indeterminate, but assuredly huge, surplus going straight to the bottom line of companies representing every sector of the economy.

* * *

By these means, capital was freed to devise accumulation strategies for application across a much-widened domain. This achievement did not unfold, however, as a straight-line reflex of capital's strength; on the contrary, it originated in the corporate state's mobilization to address capital's growing economic vulnerability. As it took hold, moreover, accelerated commodification has fostered instabilities and contradictions as well as profits.

Information and the Crisis

In 1984, Herbert I. Schiller observed, "The likelihood of achieving a stable world system—the prerequisite for the successful operation of an information-based global order under transnational corporate direction—recedes measurably with each initiative in the construction of that order."[65] A brief inventory of relevant recent trends confirms the prescience of this forecast.

One crucial change was that the world political economy became increasingly multipolar, in the lead-edge area of information and communications as well as in more traditional fields, notably, manufacturing. Capital's international embrace of accelerated commodification may have aided profits, therefore, but it did not restabilize U.S. dominance. Instead, it led to heightened intercapitalist competition. Formidable transnational suppliers took root in Europe (Nokia, SAP, Telefonica, Reed-Elsevier), Japan (Fujitsu, NEC, Hitachi, Matsushita, Sony, DoCoMo) and—as we will see later in this book—also in some less-favored areas. Non-U.S. capital even acquired the mantle of leadership in selected information and communications technology markets. To the surprise and chagrin of U.S. policymakers, for example, in 2002 Japan's NEC claimed momentary technological supremacy in supercomputing.[66] Nokia had already overpowered Motorola in global sales of cell phones. Such contests continued. In 2005, "in a direct challenge to rival U.S. satellite manufactures," Europe's top satellite makers, EADS and Alcatel, joined with government space agencies, hoping "to create a new generation of high-powered spacecraft featuring the latest onboard signal-processing technology."[67] The erosion of U.S. dominance over science-based industry was widely recognized and assailed by corporate captains of the information industry.[68]

A second trend, meanwhile, was that uneven economic development took new forms. By the mid-1990s, stagnation had not disappeared in the seven leading industrialized countries, where the number of unemployed nearly doubled between 1979 and 1995, from thirteen to twenty-four million.[69] Debt crises and structural adjustment programs continued to ravage scores of other distressed nations throughout the global South. Meanwhile, some East Asian countries—above all, China—experienced strong growth. In the United States, too, the clouds seemed to lift. Never letting their eyes stray too far from the rocketing stock market, cheerleaders trumpeted that the Internet constituted the epicenter of a "New Economy." "The business of America is . . . information,"[70] expostulated one cheerleading newspaper. Before the Internet frenzy reached its apogee, however, in 1997–98 an Asian financial crisis boiled up, seemingly out of nowhere; and the world economy again began to stagger. As high-tech-stock speculation careened to a stop midway through 2000, and the pundits shifted from talking about a "long boom" to a "burst bubble," the business cycle reached capitalism's industrial frontier. A devastating slump hit the very industries that were being heralded as the New Economy's poster-child.

Ironically, capital's success in extricating itself from an earlier round of stagnation by opening communications and information to corporate-commercial investment laid the basis for this outcome. Confirming that accelerated commodification was not a sectoral phenomenon but a more general political-economic metamorphosis, the popping of the Internet bubble triggered a wider slowdown.

Robert Brenner explains that the underlying problem lay in the recurrence—on an expanded scale—of the stagnation that had emerged during the late 1960s.[71] The scourge was again overproduction, not only in networking hardware and software but in every area from aircraft manufacturing to airline seats, papermaking and steel and automobile manufacturing to cell phones and PCs and fast food.[72] Again, overproduction was not confined to the domestic U.S. economy. The *Wall Street Journal* declared in 2003 that "countries everywhere are struggling to reduce excess capacity."[73]

How deep does the problem go? Europe continues to experience chronic

stress, while in 2005 Japan was only beginning to put behind it years of deflation.[74] The United States remains afloat only by virtue of intensified consumerism and deficit spending, sustained by what is in effect a trillion-dollar loan
from Japan and China.[75] Absent this hot money, about which first private
investors, then central banks grew anxious,[76] huge U.S. current account and
federal budget deficits, against a background of unremitting financial speculation, pose unpredictable—but grave—risks.[77]

Seemingly, therefore, the question reverts to what it was in 1970: Where
to find new sources of market dynamism? Specifically, might a new surge of
accelerated information commodification again act as a rejuvenating pole
of growth for capital?

This is not an adequate formulation of the question. Whether accelerated
information commodification continues to harbor the prospect of profitable
growth or whether it turns out to be a spent force, two additional aspects
of structural change within the contemporary political economy require
consideration.

The first is, Who will lead and recoup the unique benefits that accrue to
leadership of the global capitalist system? The political economy's growing multipolarity in information and in general has not rendered states less
important; far from it. The very fact that transnational capital's production, distribution, and marketing activities now vastly exceed the bounds
of even the largest national markets ensures that states must continue to
play a crucial role in coordinating its operation and growth—setting the
ground rules for accumulation, as Ellen Wood puts it.[78] But multipolarity
has contributed to placing new strains on the interstate system.

Though its potential vulnerability has not yet been exposed, let alone
exploited, U.S. international dominance is not permanently guaranteed.
Helping to render the deteriorating position of the United States opaque are
the concurrent economic difficulties of Western Europe and the continuing
political deference exhibited by Japan. Moreover, the United States' high-tech
capabilities, the size of its domestic market, and its global military reach still
vastly exceed those of any other nation. U.S. policymakers have responded
to multipolar competition by pumping up research-and-development and
weapons spending to apocalyptic levels and have continued to exact tribute
from the rest of the world.[79] The impression thus remains that U.S. primacy
is secure; indeed, declarations abound that U.S. supremacy has never before
reached such heights. But such exhortations must be placed in the context of
an undeniable economic decline. The United States has ballooned into the
most indebted nation in world history, while many of its blue-chip corpora-

tions have capsized in the storm of transnational capitalist competition. As U.S. economic imbalances and instabilities persist, however, aspiring rival states are beginning to emerge from the wings. Might growing economic multipolarity attain a political expression? It is no longer simply fanciful to imagine a challenge to U.S. global hegemony issuing from within the capitalist interstate system.[80] A day may come when a segment of an increasingly transnational capitalist class may choose—likely compelled by crisis—to throw its weight behind a different state or consortium of states.

In the second place, whatever economic dynamism may be triggered by a further round of accelerated commodification must be weighed against a record of harshly regressive social changes. Several factors help account for this debacle. A ferocious political offensive on behalf of capital has been ongoing for thirty-five years. President Reagan's 1981 decision to fire thousands of striking air-traffic controllers declared an open season against collective bargaining rights; the decline of trade unions has intertwined with the rise of anti-union high-tech employers, such as IBM and MCI on the supply and Wal-Mart on the demand side. While huge tax cuts benefit corporations and wealthy individuals, they also appear to justify cutbacks in government support for education, public health, unemployment insurance, and environmental protection. The results are environmental despoliation, threatening epidemics, and worsening poverty and inequality. These results are compounded by the state's swelling military expenditures, as the United States attempts to renew its global paramountcy. Alongside the other grim consequences of surging U.S. spending on war and repression have been flagrant and covert attacks on democratic ideals and practices: "blowback" from a corrupting imperialism comprises one of "the sorrows of empire."[81]

Though typically unrecognized, changing property relations in information are profoundly implicated in this wider social and political regression. Property rights in information, as in other industries, depend on the ability to exclude and penalize unauthorized—that is, nonpaying—users and forms of access, as Michael Perelman explains:

> We know that property rights require the exclusion of others from accessing property without the consent of the owner. But how do the owners of informational property rights keep others from accessing their information? . . . [P]rotection of the commodity status of information requires more intrusive protection of property rights. . . . We can be certain that the police powers of the information economy will be stronger than anything we have yet experienced. Consequently, we can assert that, given the class structure of our economy with its highly unequal division of property, including intellectual

property, the information technologies, which have an enormous potential to expand our freedom, will be applied in ways that seriously diminish our actual freedoms.[82]

Consider the technological transformation of communications and information processing. Digital media, from DVD burners, MP3 players, personal computers, and mobile phones to file sharing and other evolving uses of the versatile and capacious Internet, offer unprecedented mechanisms for creating and exchanging information. In economists' terms, these systems have greatly enlarged our capacity to develop information as a public good. Has the institutional response to these technical potentialities been to maximize and build on the newfound ability to enlarge the scope of information as shared resource?

Hardly. As capital has grown ever more dependent on information and information systems, it has mobilized to guard the emergent forms of property. According to one 1998 tally, not only Hollywood but over 80 percent of Fortune 1000 companies overall "are victims of copyright and trademark abuses on the Internet."[83] During the early 1980s, there existed perhaps twenty-five U.S. companies in the shredding business; by 2002, after curiosity was piqued by the Enron scandal, a journalist was moved to find out that they had mushroomed to around six hundred, and that a trade group—the incredibly named National Association for Information Destruction—now caters to the policy needs of corporate data trashers.[84]

This might be amusing, were it not one of many disturbing signs that exclusionary corporate control over information as private property is predicated on interweaving police powers throughout the tissue of social life. "[T]o find illicit activities it is necessary . . . to examine all digital behavior," writes one knowledgeable observer.[85] In turn, nonproprietary and uncompensated sharing often have been relabeled as criminal actions, and novel categories of felonious behavior have been proscribed. The U.S. Digital Millennium Copyright Act, for instance, makes it a crime to use or circulate technology that could circumvent systems used to "protect" intellectual property. The Supreme Court's 2005 ruling that file-sharing companies may be held liable for copyright infringement furnishes another example, as does the rapidly growing practice of so-called "digital rights management." A publication of the American Library Association explains that digital rights management employs a "range of technological tools and strategies aimed at restricting the ease with which the resources collected and maintained by libraries can be used, circulated, excerpted, and reproduced."[86] Libraries are a bellwether of the general condition of democratic information provision; their assimilation of digital rights management systems thus possesses ominous wider implications.

To police a society in which information is treated as private property, capital needs more comprehensive mechanisms of oversight and control. Corporate security requirements, symptomatically, mandate ever-increasing reliance by employees on computer passwords; one 2005 survey found that nearly one-quarter of respondents had to enter nine or more passwords to use computers in the workplace.[87] Ever-expanding corporate and government databases, together with technologies for watching, from global-positioning-system-equipped mobile phones to radio-frequency identification tags and new data-mining techniques, testify to the proliferation of invasive and extensive mechanisms of surveillance. Much is made of new vulnerabilities, as corporate media label viruses, worms, and hacker incursions a major threat to network systems and thereby attempt to justify the response—escalating control measures—as ostensibly in the general interest.[88]

The "War on Terrorism," as Robert O'Harrow Jr. shows, provides new ideological cover for elevating the threat of security breaches into a convenient general pretext for repression. Overstepping prior restraints on exactly this type of transgression, state agencies either have empowered themselves or contracted with obliging corporations to access huge pools of electronic data to track individuals through their daily rounds.[89] New mechanisms have been established for public-private coordination, and arcane new technologies for access control, system integrity, cryptography, audit and monitoring, and configuration management and assurance have been devised "to protect the computer systems that support our nation's critical infrastructures."[90]

As I write, disclosure of a vast domestic surveillance program centered in the National Security Agency suggests that corporate-state cooperation is being rapidly institutionalized. "Senior government officials," reported the *New York Times,* "arranged with top officials of some of the nation's largest telecommunications companies to gain access to important switches that act as gateways at the borders between the United States' communications networks and the international networks."[91] Grants of access to national and international telephone and Internet traffic, however, seem unlikely to have been a by-product of the 2001 attacks. Under the aegis of the National Security Telecommunications Advisory Committee, annual meetings have been held for years between government intelligence officials and leading corporate telecommunications and Internet executives. So, too, the effects of the Communications Assistance for Law Enforcement Act of 1994 (CALEA) have taken shape over a protracted interval. CALEA mandated creation of an eavesdropping capability for government over every new telephone technology, and triggered what the *New York Times* calls "a thriving 'lawful intercept' industry for technology to make eavesdropping easier." Commercial conferences devoted to the technology and practice of systematic snooping are mounted regularly, and a trade association—the Global Lawful Intercept Industry Forum—has formed. The forum's president, Anthony M. Rutkowski, claims, "I don't know of a vendor anywhere that hasn't built intercept capability into its equipment."[92]

Comparatively early on in the era of accelerated commodification, it could be seen that the political economy was diverging sharply from the forecasts of postindustrial theorists. Their prognostications bore the traces of the more benign welfare state that then governed. This divergence was well-captured by one critic in 1981:

> The stagnation in the world economy in the last few years has changed rapidly and dramatically the context of the discussion. The expectations of increased social services and education, the high and unfounded hopes for the autonomy and social contributions of a "new class" of mind workers are being set aside if not buried, with few signs of embarrassment.
>
> Wage cuts and freezes, speedup, packaged as "productivity," unemployment, the emergence of a politics using the language of "moderation" at the same time as it is slashing the living standards of the population and financing an almost out-of-control armaments expansion, and the intensification of intercapitalist rivalries, are the features of the early 1980s, likely to be extended indefinitely.[93]

Indeed they have been; moreover, the era of accelerated commodification has generated new vulnerabilities. In addition to having to defend a world system of power, U.S. elites also must oversee and police a qualitatively enlarged terrain of private property in information. Faced with mounting economic instability and financial strain and prospective geopolitical threats to U.S. world power, elites' room for maneuver has been cut. Crisis management therefore continues to be their main recourse.

By 2005, elites were divided in particular over what to do about unprecedented U.S. deficits, and authoritarian "solutions" commanded at least some open expressions of support. Peter G. Peterson, chair of the Blackstone Group (a leading private equity investment company) and of the Council on Foreign Relations, declared that, in respect of the mounting U.S. debt crisis, "We have reached a point in this consumption-obsessed country where if we want people to save, we're going to have to make it mandatory. This, I know, is very controversial. We're going to have to do what Singapore and Chile and Australia did."[94] "We must change policy direction," agreed Robert E. Rubin, Treasury Secretary under President Clinton and a director of Citicorp, because "the fiscal and entitlement holes are now so deep": "Everything should be on the table."[95] That such powerful figures can publicly voice these views—which, boiled down, seek to compel the working population of the United States to undergo a savage "structural adjustment," comparable to those imposed on numerous less-developed countries by the International Monetary Fund and the World Bank—is suggestive. One of the things that it suggests is that informationalized capitalism carries a strengthening impulse toward a full-blown authoritarianism. As policymakers grant public and private power broad license to fuse, the historical tension between capitalism and democracy appears to be in danger of metamorphosing into a stark opposition.

But the prospect before us contains more than any mere top-down will-to-power.

Intensifying Social and Political Struggle over Informationalized Capitalism

Informationalized capitalism has been neither completely nor securely established. A harsh regimen, it remains open to contestation and change. Social struggles not only persist, they too have become more informationally focused and centered. Desire and necessity, dreams and duress, have been imbued with informational elements. The hope of developing information and culture as shared resources essential to democratic reconstruction has fired diverse initiatives, while the regressive character of informationalized capitalism combines with escalating intercapitalist rivalries to render likely sharpening conflicts over information and culture. A summary inventory gives some sense of the range and variety of contemporary struggles over information.

Probably the most visible struggle within the U.S. has been over "media reform." A popular mobilization to roll back corporate-state domination of the channels of public expression, which have been continually diminished by economic concentration even as they were also clogged by a deluge of commercial and government propaganda, has gained momentum. Of equal import has been the growth—and not only in the United States—of oppositional and community media, employing technologies from microradio to cable television, wireless, and the Internet.[96]

There is also an urgent need, and significant public inclination, to push back against unregulated corporate data-brokers like Lexis-Nexis, Choice-point, chain retailers, banks, and others.[97] It is not simply an individual right to privacy that is being subverted, but the citizenry's collective democratic rights. Data protection, further advanced in Europe than in the United States and claiming radical and conservative adherents, has become a politically important priority.

Internet governance represents a third area of struggle. The notion that there is no such governance because the Internet empowers the grassroots is increasingly being seen as a delusion. At the U.N.'s World Summit on the Information Society, a series of deliberations between 2003 and the end of 2005, no action-item was of greater importance than Internet governance.[98] Criticisms of the existing system stress, aptly, that the interests of trademark owners have been elevated above those of ordinary users; that the existing administrative mechanism is opaque and unresponsive; and that the U.S. government controls key functions, through its carefully camouflaged institutional power over the domain-name system, while other governments have been substantially excluded, so that the "universal jurisdiction" of the Internet is effectively administered not by the international community but by the United States.[99] Perhaps remembering "the U.S. presidential threat in spring of 1980 to deny Iran access to international communications systems, including Intelsat,"[100] in 2005 some nations (Brazil was mentioned) "expressed concern that the U.S. government, acting through ICANN, could unilaterally . . . cut them off from the Internet."[101] Internet governance is destined to become more controversial during the years ahead, as the functions of the Internet multiply within social life.

Private-property rights in information present another sweeping issue that encompasses a series of struggles undertaken by community activists, academics, technical workers, and reform groups and individuals on behalf of freer information circulation and nonproprietary media systems. Fights for the public domain unfolded around an impressive array of sites and under different rubrics: "free culture," a "creative commons," a "public library of science," certain kinds of open-source software, a "cultural exception" to World Trade Organization rules on audiovisual works (see chapter 6), and unfettered access to cheap medicines. The common element has been the struggle against continuing enclosures of nonproprietary information. There is every reason to think that these seemingly desperate efforts will aim for—and find—increasing shared purpose and organizational resolve.

Though it became less visible as a policy issue, owing mostly to public-relations smog, the digital divide continues to animate struggles across the world. The relevant issues go far beyond unequal social access to hardware such as PCs and networks. They encompass more than disparities in access to software and services as a consequence of income inequality, unevenly distributed computer-literacy skills, and the overwhelmingly disproportionate reliance of the Web on English (and, secondarily, perhaps a dozen other languages). The digital divide is, most profoundly, about the distribution of social power to make policy for the production and distribution of information resources. Unless that power is broadly shared, democracy itself is threatened.

To take up these challenges and to renew the struggle for a New International Information Order effectively requires careful and sustained engagement with the political-economic changes that have been reordering information and culture. The remainder of this book contributes further to that engagement.

PART 2

Vectors of Commodification

4

Business Users and U.S. Telecommunications- System Development

Visionary executives, brilliant engineers, and even regulators have been credited with charting the development of the U.S. telecommunications system, while business users of network systems and services remain faceless and scarcely noticed as a shaping force.[1] However, down through the decades, industrial, financial, and commercial telecommunications users have played a formative and even a determining role in the evolution of this vital infrastructure. They have accomplished this through a combination of private organizational action and public political activism. Their growing reliance on networks and their rising stake in policy debates—especially over the integration of new networking technologies—have motivated business users to organize, so as to project their interests and needs more effectively. By engaging this vital but little-known history,[2] this chapter lays groundwork for one of this book's major themes: informationalized capitalism constitutes a general rather than merely a sectoral political-economic phenomenon.

Business Users and Early System Development

The history of telecommunications has been written largely from the supply side, to stress the role of the carriers that provide society with access to different network systems and services. This has been unfortunate as well as mistaken, because it has recurrently helped to make antimonopoly principles—rather than social need—the preferred framework for policy.

From the beginning of the American republic, the demand side of the telecommunications equation has been pivotal; business users have repeatedly made critical contributions to the structure and policy of telecommunications. Prospective business-user demand helped justify rapid expansion of an increasingly encompassing post-office system throughout the early national period and generated momentum for continuing technical and organizational innovation in the supply of network services throughout the late nineteenth century.[3] The fledgling telegraph was disproportionately used by large-scale enterprises oriented toward a truly national political economy: banks, commodity traders, news agencies, and railroads.[4] Telephony too originated largely as a business—or, better, a business-class—service.[5] Service applications were unevenly generalized across the field of big business, which came to prominence late in the nineteenth century in part through the innovation of networked business processes.

Like a privileged minority of individual users, businesses employed the public switched telecommunications network; unlike individuals, however, businesses also came to depend on a category of service aimed solely at the organizational market: leased lines, or private wires, as the specialized facilities proffered by carriers beginning with Western Union are called. By 1878, private wire contracts for Morse service had proliferated to the point that one vendor specializing in service to banks and brokerages rented three hundred private lines with twelve hundred miles of wire in and around New York City.[6] The Bell telephone system likewise quickly moved into leased-line markets.[7] Provision of this specialized class of service—"the greatest source of revenue of the telegraph companies," reported the *Wall Street Journal* in 1909—was lucrative.[8] By one account, 31 percent of Western Union's net income derived from "sources other than toll messages" in 1896, but the proportion ranged upwards of 60 percent by 1906–8.[9] A Western Union official stated to a congressional panel (reporting in 1909) "that this business was so much more profitable than handling messages that the company had considered a suggestion that it cease to handle messages entirely and turn its entire attention to leased-wire business."[10]

Business users of leased lines included brokers, which possessed systems "covering the entire country." Apart from the railroads, however, Standard Oil was the largest individual business user, followed by "the packers," and then U.S. Steel, all of whose plants and subsidiaries, a newspaper reported, "are connected by private telegraph wire." The Associated Press was another heavy user of leased lines.[11]

As they increased their reliance on telecommunications in general and leased lines in particular, business users began to organize themselves into a pressure group of rising importance to the structure and policy of telecommunications provision. Early interventions proceeded on an ad hoc basis. During 1904–5, for example, private consultations were undertaken between New York Telephone, the Bell affiliate, and a Telephone Committee established by the Merchants' Association of New York to conduct "an exhaustive examination" of telephone service and charges there. The committee stated in its report that New York Telephone had obligingly "consent[ed] to open its books and to supply the Committee with all necessary details of investment, gross earnings, operating expenses, and net earnings." The immediate result was a series of rate adjustments and service changes.[12] But the more important consequences were the beginnings of formalizing a policy-making role for business users and stabilizing a shaky and inadequate telecommunications system.

Facilities-based competition in telephone-service provision had erupted unevenly throughout most of the United States in the years following the lapse of the Bell patents in 1894, so that by 1907 around half of all subscribers accessed networks operated by non-Bell carriers. The Merchants' Committee, significantly, faulted this competitive system for furnishing an inadequate and overpriced service.[13] Business users thus implied their willingness to support a bid by Bell to reestablish its monopoly, if in return the telephone trust promised to acquiesce to legally structured rate regulation and, equally important in the committee's view, "consistent and reasonable publicity" for supplier charges and services.[14]

Given the importance of New York City–based businesses to the entire country's finance, commerce, and manufacturing, this policy choice reverberated widely, and business users possessed of similar intent were also organizing elsewhere.[15] In opposition to continued concerns on the part of trade unionists, political reformers, and other groups, the committee's rejection of competition helped, as Alan Stone suggests, to shift "elite opinion . . . definitively in the direction of local monopoly's provision of telephone service" and ultimately toward what Stone calls the "regulated network manager system" that AT&T thereupon succeeded in dominating once more.[16]

A decade later, the role of business telecommunications users began to extend toward international policy objectives as well. U.S. diplomats hoped to use a post–World War I conference on international electrical communications as a basis for projecting U.S. power more effectively in this increasingly critical sphere. The National Foreign Trade Council, representing U.S. exporters, overseas traders, and executives, wrote to the Department of State in November 1919 to demand that business interests be expressly represented in these deliberations. Other leading trade associations, such as the American Manufacturing Export Association, the Merchants' Association of New York, the National Association of Manufacturers, and the U.S. Chamber of Commerce, met before the conference "to develop a unified position in international communications." They succeeded not only in agreeing upon an agenda but also in pressing it on U.S. diplomats, who altered their negotiating positions to accommodate users' concerns.[17]

It is therefore correct to claim—in contrast to those who would lay this achievement at the door of AT&T executives—a critical role for business users in establishing the system of U.S. national and international telecommunications that began to cohere in the years around World War I and was consolidated, with significant alterations, throughout subsequent decades.

Radio Technology and the New Deal Settlement

The next great watershed was reached around the technology of wireless signal transmission. The story of radio has been told often, but, since pathbreaking but now largely forgotten research by Murray Edelman and Dallas Smythe, it is relayed without concern for the nonbroadcast, non-common-carrier services staked out and developed by business users.[18] This omission is significant. Business users of telecommunications in fact converged on radio technology during the 1920s and advanced unprecedented claims on the still-forming nationwide system, especially in regard to the high-frequency, short-wave spectrum band that was beginning to be exploited commercially.

The Federal Radio Commission (FRC), established in 1927, made available for assignment portions of this new band (extending between six thousand and twenty-three thousand kilocycles), reporting that "a constantly increasing number of applications for the use of these frequencies has flooded the commission, covering a wide variety of services and experiments." As a result of this discrepancy between available supply and demand, the agency determined to undertake an extensive investigation of the properties of the high-frequency band, "the needs and merits of the types of service seeking accommodation in the band, and the application of the standard of 'public interest, convenience, or necessity' to these questions." Absent such "a scientific and orderly plan" for the exploitation of these higher frequencies, the commission explained, "congestion equal to that which has been the root of all evils in the broadcast band would obtain" here as well.[19]

A public hearing on these issues was held early in 1928. In attendance were all of the major executive agencies with a direct and indirect interest in communications, from the Departments of State, War, Navy, and Commerce to the Bureau of Lighthouses. Also represented, as scholars have long recognized, were the leading radio manufacturers and the domestic and transoceanic communication companies. But the range of organizations touched by high-frequency-allocation policy extended beyond these familiar faces. The FRC observed that the following groups, "represented in many cases by eminent radio engineers and lawyers, were called upon in turn and each made an earnest plea for accommodation in the high-frequency band":

Newspaper services. . . .
Airplane operating companies.
Navigation companies.
Railroads.
Department-store chains.
Electric railways.
Interurban bus systems.
Electric power transmission systems.
Lumber companies. . . .
Motion picture producers.
Police and fire-alarm systems.
Forest and watershed patrols.
Ranch owners.
Remote resorts and hotels.
Operators of facsimile transmission services. . . .
Mining and oil companies.
Packers and shippers.
Geologists.

Disparate organizations and associations of business users presented them-selves: the American Petroleum Institute; Firestone Tire and Rubber and its rival, Goodyear; the retailers R. H. Macey and Bamberger Company; media companies including Universal Pictures, Hearst Publications, McGraw-Hill, the *Los Angeles Times,* the *San Francisco Examiner,* the *New York Times,* the *Chicago Tribune,* and other newspapers; and the American Railway Association. Marked disagreement ensued, as rivals vied for spectrum with which to satisfy "such strikingly different services as transoceanic and transcontinental communication, railroad needs for communication between locomotive and caboose on a freight train and between office and switch engine, the claims of oil companies not only for communication purposes but also for prospecting for oil, and of power companies for emergency purposes." The welter of com-peting interests and demands was such that the FRC determined to allocate frequencies at first only in what it termed the transoceanic high-frequency band, "in order that these frequencies should not be appropriated by other nations to the disadvantage of the United States."[20]

While unhappy would-be users pursued their cases in the Court of Appeals, the following year the commission got started on allocation of the so-called continental high-frequency band (1,500 to 6,000 kilocycles). Applications were made by "several large concerns desiring to establish public systems of point-to-point radio communication in the United States, duplicating the wire systems between the larger cities." Also considered, however, were "a large number of applications from more or less private interests desiring to set up a more limited system of communication, such as between chain stores, brokers' offices, mail-order houses and their branches, oil companies, mines, and the like."[21] And industrial demand for radio frequencies increased. The FRC reported in 1929 that applications had been entered "for the assignment of literally thousands of frequencies more than are available."[22] It promised weakly to bring "[t]he best engineering talent in the country" to bear on the policy issues it confronted.[23]

In retrospect, we can see that a varied group of business users was beginning to formulate a qualitatively new demand: access to specialized communication facilities and services, to be integrated into large business operations as a private matter, outside the sphere of common-carrier (or broadcast) provision. Protesting to legislators that the established carriers remained inert, unwilling or unable to meet their dynamic telecommunications needs at a suitable price, business users also were beginning to organize themselves into a permanent national pressure group and policy-making force.

In 1929, the FRC responded by granting specific groups of business users limited rights to develop specialized network services. This support was selective and carefully delimited: it did not yet impinge centrally on or require reorganization of the nation's public switched telecommunications network. But it did sustain burgeoning initiatives in the networking of productive, distributive, and commercial activities by major corporations and industry consortia. We possess almost no historical research into these proprietary efforts, each of which expressed and built toward a specific industry's efforts to integrate telecommunications into its workaday operations. The geophysical exploration and pipeline services innovated by oil companies comprise one case in point; frequencies were first set aside for geophysical service in 1930, and by 1947 five hundred stations and forty-nine frequencies had been licensed to oil companies to help them locate oil deposits.[24] A second case coalesced around the shared radio services developed by Aeronautical Radio, a consortium of airline companies, beginning in 1929–30.[25] During the 1930s and 1940s, railroads, utilities, motion picture companies, and newspapers were allotted radio frequencies with which they would build out specialized intra- and interorganizational business services.

It is in this context that AT&T's mid-1920s deal with fledgling network radio broadcasters—which granted to the telephone company a monopoly over radio program transmission in exchange for its promise to withdraw from radio broadcasting markets—may be seen as a privately negotiated attempt to avert development of such special-purpose industrial radio systems. Radio broadcast networks and the television networks that succeeded them in this instance agreed (in the latter case, under duress) to contract for private line service from AT&T rather than build their own proprietary networks.

For the most part, U.S. industries followed this course and leased lines from the telephone company. During the mid-1930s, AT&T furnished around a thousand private line systems to business users, including 250 used by banks and financial houses; in 1936, private line toll services supplied $32,590,337 in revenue to AT&T—10.5 percent of its overall intra- and interstate toll revenues and around one-quarter of its interstate long lines revenues.[26] In New York City at this time, private branch exchange attendants—women workers operating switchboards in hotels, large offices, hospitals, stores, and other businesses—outnumbered telephone company operators three to one.[27]

Business users thus attained an expanding political and economic status in telecommunications, mostly—though not exclusively—within the context of the regulated common-carrier system over which AT&T and, residually, Western Union presided. Business users' needs were relegated to the edges of the evolving public switched network as a result of deliberate and considered public policy.

This was partly because the extent of industrial demand for frequencies was still relatively slight; networked business processes were only beginning to proliferate. A second factor was the palpable scarcity of the enabling resource: usable electromagnetic frequencies. However, even before the New Deal period, the FRC policy of placing business users' demands for access to radio frequencies after those of the common carriers expressed other important elements. "Those applicants proposing to engage in the communication business serving the entire public or a particular class of the entire public, and assuming the duties, obligations, and responsibilities of common carriers, are deemed to be in a better position to meet the standard of public interest than any of the other applicants," was the commission's opinion in one precedent-setting early case.[28] In another decision, the agency expressly held "that applications would not be granted for service which would duplicate that already furnished by land-line companies." The reasoning employed in support of this finding—and the policy from which it followed—merits extended quotation:

It may be that the commission owes the wire telegraph companies no duty to protect them from competition by radio services. But there is a much broader consideration than this. The commission, while encouraging the development of radio, should nevertheless, in applying the statutory standard, take into consideration the possibility of a radio company competing unfairly with a wire service to such an extent that the general public may suffer. . . .

Obviously there is no constant relationship between the capital, personnel, and maintenance expenses of a wire circuit on the one hand and its volume of traffic on the other. The company's cost of a wire circuit between small communities is not always justified by the income from traffic. The offices in small communities must be maintained to preserve the utility of the entire service to all the people of the Nation. The charges for message traffic over the more profitable circuits between large centers of population must include some charge for the maintenance of the less profitable circuits. The wire companies' charges for their readiness to serve are thus equitably distributed.

With the wire communication companies thus situated, the commission can not, from the standpoint of national welfare, encourage the establishment of radio communication systems based solely upon the selection of the most profitable points of communication. Radio companies taking the "cream" of the business at reduced rates might impair the utility and the economic structure of the wire companies, for the latter, in order to meet competition, might be compelled to abandon unprofitable circuits. . . .

Upon the same considerations, the commission must not lend itself to the establishment of radio circuits which will rely upon the handling at reduced rates of the bulk traffic of individual large corporations between their various offices, to the practical exclusion of the less profitable occasional traffic of the general public, especially under circumstances where the wire communication companies are prevented by law or regulation from making such preferential and discriminatory arrangements.[29]

Scarcity in a context of skewed and uneven demand, disproportionately dominated by business users and urban locations, required that the national system—the public switched network—be assigned priority in spectrum allocation and system development. This conception would be strengthened as a consequence of additional New Deal reforms. More immediately, however, the FRC's attempts to integrate radio technology coherently into what was becoming a multifaceted telecommunications system suggested a need to unify the existing fractionated federal regulatory mechanism. The Interstate Commerce Commission exercised nominal jurisdiction over nationwide telephony and telegraphy; the Federal Radio Commission oversaw radio. As the quote above makes clear, however, convergence between radio and wireline systems demonstrated that both needed to be encompassed for policy-making purposes by a single regulator.

Numerous issues were raised, and many parties represented, in the attempt to create a Federal Communications Commission (FCC). Again, however, what stands out in prevailing accounts is the omission of business users—and again, this slight is unwarranted. In the context of deep economic depression and significant technological change, influential business users inserted themselves into the endeavor to establish a more far-reaching communications regulator. Lacking effective rate regulation for international radiotelegraphy, a "cable and radio users' protective committee" insisted in congressional hearings on the need for a Federal Communications Commission with rate-setting authority. These large users declared that the carriers were colluding with the British Post Office to extort higher charges from them.

Rate reductions had been instituted earlier, as shortwave radio and loaded permalloy cables were placed into use and overcapacity in turn developed on North Atlantic routes; only by preserving these lower rates could the very large telecommunications expenses borne by the complainants—some fifty-odd banks, stock and commodity exchanges, and import/export houses—be kept within restraint. But the carriers instead had ostensibly combined in hopes of enforcing a mutually beneficial rate *increase*. In these depressed times, such a predatory imposition was especially onerous. "Unfortunately there exists in this country," wrote the chairman of the Cable and Radio Users Protective Committee to the chair of the Senate Interstate Commerce Committee, Clarence C. Dill, "no tribunal with adequate powers to consider and determine this conflict of opinion between the companies and their customers." "That such a body be promptly established," the user group concluded, "is our only request."[30] Urged on by the State Department, Congress passed the enabling legislation, and, to the chagrin of AT&T executives (who opposed it), the FCC was established.

In the context of the Great Depression, which brought the question of how to overcome economic stagnation to the forefront of policy making, emboldened New Deal regulators began a top-to-bottom review of the process of telecommunications-system development.[31] Through the so-called Telephone Investigation of 1935–39, they began by taking a hard look at the internal economic workings of the AT&T monopoly. In that process, they established an important forerunner of the sweeping investigations of U.S. industry conducted by the Temporary National Economic Committee beginning in 1939. Substantially altered policies followed hard on the heels of this unprecedented scrutiny.

Though problems of jurisdiction frequently erupted, the FCC and existing state public-utility commissions sought to develop an effective partnership; only through such cooperation could they also forge a substantive capacity for meaningful, end-to-end oversight of the nation's telecommunications network. The new framework's limits narrowed as the New Deal gave way to the cold war, but regulators nonetheless succeeded in reorienting the system so as to place it on a limited public-service basis. By providing government loans to build out the network and artfully rebalancing local and long-distance rates, regulators and legislators boosted residential demand. During the long boom that followed World War II, the chronic undersupply of local residential telephone service at last was relieved, not only in the city but eventually also in rural districts. Inclusive or "universal" household access to the telephone thus became the signal achievement of the New Deal's public-service settlement. But other policy changes were also made.

More encompassing labor reforms, for example, impinged importantly on telecommunications. By establishing employee associations and introducing other welfare-capitalist reforms, for a generation after World War I, AT&T executives held independent trade unions at bay. However, between passage of the National Labor Relations Act in 1935, the ascent of the Left-led CIO unions, the American Communications Association, and the United Electrical Workers during the decade that followed, and the formation out of a massive strike in 1947 of the more conservative Communications Workers of America, the principle of collective bargaining rights by independent unions took firm root.[32] A separate but also vital policy reform opened up AT&T's intellectual property. A federal antitrust case brought against the carrier in 1949 eventually resulted in restricting AT&T's field of endeavor to regulated telephone service markets, while subjecting its Bell Laboratories unit—the country's preeminent industrial research operation—to a new regimen of compulsory patent sharing.

Public service was not all it was cracked up to be; there were substantial areas of practice into which it was never extended. Gender and race discrimination in telephone-industry employment practices were one such field. The power of the Pentagon over telecommunications policy was another. The fact that regulators remained frustrated in their attempts to exercise real oversight of AT&T's unregulated manufacturing subsidiary, Western Electric, constituted a third prominent limit on public service. The FCC's unbroken support for what it began to term "industrial" radio applicants denoted a final, increasingly vital, restriction over common-carrier system development.[33]

At first, the need seemed to the FCC to be merely to formulate requirements and make recommendations pertaining to "the increased demands for frequency use for all purposes which would follow the conclusion of the war."[34] Dampened demand by businesses for special-purpose networks during the Depression, followed by the overarching focus on wartime communications requirements, had permitted the agency to defer policy making in this area. But the FCC's premonition that it would have to grant more attention to the needs of industrial telecommunications users turned out to be insufficient to the task at hand. Spiraling business demand and hothouse network technology development during the cold war would compel the FCC repeatedly to revisit the question of specialized communications systems for corporate users to help integrate another and even more fundamental innovation: the electronic digital computer.

In this drawn-out process of policy change, the New Deal emphasis on public-service telecommunications was placed under increasingly fierce pressure. Technological revolution around networks thus occurred by way of increasing political regression.

Reactionary Modernization: Business Users and Computer Communications

World War II and the cold-war mobilization that followed it powerfully spurred innovation in electronics, telecommunications, computers, and aerospace. System development across the field of networking gained a more discrete, though less well-known, boost from the settlement in the late 1950s of antitrust cases against AT&T, IBM, and RCA.[35] The Justice Department thereby succeeded in what Stanley N. Barnes, its antitrust chief, called a "program to open up the electronics field" by compelling AT&T, IBM, and RCA to license their patents on easy terms.[36]

Building not only on "cost-plus" contracts for cutting-edge computing technologies from military agencies but increasingly on demand for cost-efficient applications by a diversifying base of industrial, commercial, and financial users of punched-card data-processing equipment, during the late 1950s and 1960s the U.S. computer industry took off. Revenues garnered by IBM—the industry leader by a long shot—for its newer electronic computer systems exceeded those from conventional punched-card systems in 1962, prompting executives to bet the company on a new generation of electronic computers (the 360 line, introduced in April 1964).[37] Again, however, it was not only conducive policies on the supply side that produced this boom. "The return to a 'peace-time' economy" after World War II "opened a flood gate of demand by industries that had been deprived of equipment during the war, and new uses and methods devised during the war further increased demand as customers applied them to peace-time activities."[38] Prosaic but vital processes began to be networked, in whole or in part, reaching functions such as payroll, personnel files, insurance records, accounting, banking, inventory control, and manufacturing production scheduling.[39]

Jump-started by the military's Project Sage, an air-defense system developed by the Massachusetts Institute of Technology in partnership with several major corporations but targeting a wide range of existing and prospective business applications, digital data transmission using telephone lines became an increasingly vital focus.[40] Business users in particular sought to develop computer communications networks to spread the benefits of their centralized data-processing resources more widely throughout their organizations; and to innovate remote-processing applications, such as the cutting-edge Sabre system developed by American Airlines in partnership with IBM.[41]

The telephone network had been engineered for voice applications, however, and did not lend itself easily to data traffic. "The voice common-carrier communications systems of today," complained Paul Baran, one of the engineers who designed the Internet's underlying technology of packet switching, in 1967, "were designed primarily to provide a voice-to-ear or typewriter-to-typewriter link between humans. Today's communications regulation doctrine still regards the computer merely as another user of these existing telephone, typewriter, and telegraph networks. It isn't merely the matter of jamming a size ten foot into a size five shoe. The fundamental desired communications characteristics are so different that we are living on a procrustean operation basis."[42]

Adapting and adjusting the nation's telecommunications system posed basic problems of public policy. Should computer networks be incorporated into a regulated, heavily unionized, public-utility telecommunications system, built around end-to-end provision of voice service and making prominent room for residential users? Under whose auspices could the vibrant and expansive field of data communications be most adequately developed? The telecommunications industry that was dominated by AT&T? Or a punched-card data-processing business that was feverishly transforming itself into an electronic digital computer industry but that was almost equally monopolized by IBM?

As we know, the answer turned out to be neither. What accounts for this outcome? Why did computer communications including, ultimately, the Internet, evolve in this curiously backhanded way?

A report written by a staff member of the FCC self-servingly avers, "[I]n providing fertile ground for the growth and development of data networks over the nation's communications infrastructure," over a period of thirty-five years beginning in the mid-1960s the FCC determined through a series of proceedings "that computer-based services offered over telecommunications facilities should not be subject to common carrier regulation"; in so doing, it asserts, "the Commission set forth the necessary unregulated landscape for the growth and development of the Internet."[43] Was the FCC a far-sighted and consistent (de)regulator?

The claim possesses nominal validity. The FCC did in fact contribute, through an episodic series of vital decisions, to the emergence of computer communications in general and the Internet in particular. But to leave the matter here is to neglect the vitally relevant fact that these same FCC decisions were predicated on and, in case after case, prompted by specific, concrete demands put to the agency by business users of telecommunications, generally in alliance with a small but rapidly growing group of independent equipment and service companies. Large business users, working together as a super lobby comprised of trade associations and newly established lobbying groups, must be accorded pride of place in explaining the modernization of U.S. telecommunications around data networks. The driving force behind computer communications was thus not simply that the FCC chose to expedite its development but that business users effectively destabilized, and ultimately supplanted, the regulated network manager system that they themselves had earlier helped to erect.

Why? By three vital standards—an essential role in coordinating increasingly dispersed corporate units at a moment of sweeping and protracted economic expansion; concurrently escalating business expenditures; and a cascade of new applications carrying vital strategic importance for further growth—during the 1950s and 1960s network systems were transforming into an ever more essential business infrastructure. The decision to intervene in a continuing forceful way in policy making occurred because corporate telecommunications users, who were already widely reliant on specialized telecommunications systems, began to demand access to something more than an expensive, voice-oriented, regulated network. As business-user dependence on merged computer communications deepened while AT&T continued to operate within the terms of the New Deal settlement, consensus over development policy for the nation's telecommunications system shattered. Disagreements developed over the specific uses to which the public network might be put, the need for special-purpose equipment and services, and the prices at which different services would be provided. Elsewhere I have traced the resulting spiral of policy change, beginning in the late 1950s; through a series of related telecommunications regulatory proceedings at the FCC, AT&T's interests were pitted against those of its biggest customers.[44] Business users demanded, and the authorities made increasing haste to provide them with, "the same latitude in the use and implementation of [their] communications facilities that [they] enjoy in the use and implementation of the many thousands of other tools, facilities, and services necessary to the conduct of [their] business."[45]

Business users probably generated a majority of the carriers' overall long-distance revenues, and a small group of large users accounted for most of this total.[46] Such massed demand lent weight to their attempts to organize into a pressure group, possessing power that was not available to small residential rate payers—power that was wielded to obtain successive policy changes from the late 1950s to the present day.[47] Deliberation over the future of U.S. telecommunications-system development, however, reached a watershed between 1967 and 1970.

An executive-branch initiative undertaken by the Democratic administration of Lyndon Johnson must be credited with a decisive role in transforming business users' demands for telecommunications liberalization into a bipartisan political consensus. A Task Force on Communications Policy convened by Johnson in 1967 drew together members of over a dozen federal departments and agencies and employed research contractors at universities and think tanks. Thus the Task Force's endorsement of limited competition in private line service and its more general prescription that telecommunications policy "should seek to develop an environment always sensitive to consumer needs" resonated widely.[48]

The Task Force's chairman, undersecretary of state Eugene V. Rostow, likewise interpreted his mandate broadly and commenced to undertake what he boasted was "the most fundamental and broad gauged study of communications policy in forty years."[49] His move to "rear back and look at communications as a whole" spurred intense disagreement among Task Force members during the fall of 1967, but Rostow prevailed, gaining authority to broaden the initially more-limited study to encompass domestic common-carrier (and broadcasting) issues.[50]

Days before concluding the Task Force's deliberations, Rostow signaled directly to Johnson that he had succeeded in crafting far-reaching proposals for policy change:

It takes a long time to get a Task Force of this kind established and into motion. The opportunity should not be wasted. In my view, the line of policy laid out . . . is moderate, balanced, and right, and its articulation in a Task Force Report could help to influence the pattern of decisions by the FCC, by industry, and by Congress in a constructive way for a long time to come. Our efforts and the discussion of the draft chapters have already had a marked effect both on the FCC and on AT&T policy.[51]

The reforms the Task Force called for included all three of the most vital initiatives concurrently under review by the FCC.

While suggesting that "integrated provision" of domestic public-message telephone service be preserved, the Task Force expressly recommended that the FCC endorse "the removal of unnecessary restraints to promote innovation and to encourage greater responsiveness to consumer needs" in the case of "services which supplement those of the basic public message telephone network."[52] The Task Force artfully labeled "teleprocessing" such a "supplemental" service, for which "the removal of tariff restrictions on the sharing of communications lines, on splitting or resale of channels, and on message switching, seems compatible with maintaining the integrity of the basic communications network." The Task Force also supported liberalized foreign-attachment provisions, without which new kinds of network equipment—preeminently, computers—could not be used in concert with telecommunications facilities.[53] A third proposed policy change was also crucial to the Task Force's agenda in domestic common-carrier telecommunications. Inaugurating what has since become a familiar litany, the Task Force attacked regulation in favor of competitive, or liberalized, market entry into telecommunicatons; it argued that the FCC should approve licenses for prospective carriers such as MCI, which had petitioned for authority to compete with AT&T in specialized business markets.[54]

With these policy changes, which were all approved by the FCC, U.S. telecommunications-system development policy was radically destabilized in two ways. On the one hand, by treating computer networking as a *supplementary* service domain, policy makers were actually creating space in which business users and specialized suppliers could work to build and assimilate powerful new networks and network applications across the length and breadth of the political economy. Through episodic extensions of this policy, the "supplementary" computer network segment was enabled to cannibalize the public-service core of the nation's telecommunications system. On the other hand, while acknowledging that newly authorized carriers "could raise serious problems for the integrated network," the Task Force also asserted that "these problems can be met by allowing the established carriers sufficient flexibility in rates to meet competition, and by strengthening regulatory capabilities to prevent destructive competition."[55] While the Task Force's caution about problems associated with "destructive competition" ultimately would be shown to have been fully warranted, its assurances that these problems could be successfully contained would not.

Business users' prime demands for system development struck at the heart of existing policies. First, they proposed that the cost-economies achieved by introducing new technology into the telephone network should be passed largely on to them instead of being directed toward other ends (such as protecting low local residential-telephone rates): acceptance of this tenet would dramatically destabilize the comprehensive rate structure that had been developed to build out inclusive network access. Even more radically, they proposed that they themselves, in partnership with emerging specialized providers, should be accorded proprietary control over a large and increasing segment of the networking process: agreement would subvert the New Deal norm of comprehensive, or end-to-end, public-utility regulation of "the" public switched network.

A renewal of the public-service conception to accommodate computer-communications networks, contrary to what actually occurred, would have inescapably reinvigorated that doctrine. What and whose needs should such an emboldened public-interest principle address? How should it be updated, and what policies should it seek to embody? Should the requirement of universal telephone service, just coming into its own in practice, be enlarged to include comprehensive public access to networked information services? Should nondiscrimination mandate comparable service for all users of computer communications? Should public-utility status be conferred on the computer industry? Such were the vital questions that the FCC bypassed because of business-user intervention in favor of devolving fundamental decisions about ownership and control of computer networks on proprietary corporate interests. Through one proceeding after another, regulatory controls were pared down and market restrictions relaxed. The process of liberalization moved from the periphery of the public switched network to its center. As ownership and control of network technologies were lifted out of the web of obligations that had encased the regulated network manager system, AT&T's own strategic thinking ultimately transformed into that of just another proprietary competitor. The federal antitrust suit that instigated the 1982–84 breakup of AT&T climaxed this trend. Curtailing the system of end-to-end general-purpose service offered by a regulated monopoly provider, as Alan Stone underlines, the divestiture concurrently signaled "the atrophy of the public service principles" that had infused the New Deal settlement.[56]

Corollary initiatives were undertaken in international telecommunications for similar reasons. Since the buildout of long-distance submarine telegraph systems during the nineteenth century, international networks had functioned as prized "tools of empire"[57] and had been used to project political-economic, cultural and military power outward on behalf of imperializing elites. Control by national corporate-state interests of supranational networks was, in turn, long deemed to be a gravely important policy objective. Vying strategically with their counterparts in Europe—above all, England—and Japan, U.S. officials and corporate executives worked zealously, from early in the twentieth century onward, to erect an American system of international telecommunications; but their efforts continued to meet with only partial success.[58] A U.S.-supplied and -controlled supranational system was finally actualized with the establishment of the International Telecommunications Satellite (Intelsat) consortium in the early 1970s. A study by the Department of Commerce summed up elite opinion that Intelsat constituted "an unqualified, outstanding success on institutional, financial, and operational grounds, and must be considered a triumph of U.S. foreign policy."[59]

Built around the state agencies that controlled national telecommunications networks throughout most of the world, Intelsat furthered the development of a global satellite system by uniting dozens of national communications ministries as signatories in a system whose major owner was a private U.S. corporation, Comsat. Through its rate-setting policies, Intelsat ensured that the reach of its emerging network would be global. Heavy-traffic routes—notably across the North Atlantic—paid more than would have been warranted by cost-based pricing in support of a single coherent network granting access to most of the world; low-density links—say, between Brazil and South Africa, or Indonesia and India—paid less.

After considerable deliberation, the Reagan administration eventually chose to jettison this U.S. achievement. Tipping the balance against Intelsat were the prospect that the U.S. monopoly over space was being broken by European and Japanese rivals and mounting demands by transnational corporate users to curtail existing national controls, through which states had been able to regulate important features of satellite service as well as pricing.[60] A third factor was also vital: new, private satellite carriers sought to become rivals to Intelsat.

Transposing the precepts of liberalized entry to the international market,

Reagan authorized competing private satellite systems, which duly proliferated alongside regional networks established by other countries and foreign
capitalists. Through unflagging U.S. effort, this liberalization mandate was
substantially extended and enlarged in the 1990s, most notably through a
1997 pact on basic telecommunications acquiesced to by members of the
World Trade Organization. National flag carriers began to be supplanted by
supranational corporate-run networks, and authorized levels of foreign direct
investment in telecommunications increased, even in the United States.[61]

U.S. negotiators were confident that, whereas U.S. carriers would pig-
gyback on the WTO agreement to snap up or build out new subsidiaries in
dozens of other countries, only a select group of foreign companies could
conceivably reciprocate by acquiring or building a nationally significant
U.S. carrier. This stance appears to have been justified. While U.S. carriers
and investors initially obtained stakes in network infrastructures within
scores of countries, the size and wealth of the U.S. market combined with
episodic government interventions to keep it mostly off-limits to foreign
capital. Just as the boom was ending, Deutsche Telekom—one of the largest
European carriers—had to pay fifty billion dollars to take over what was then
the eighth-largest U.S. cellular company, VoiceStream Wireless (renaming it
T-Mobile).[62] However, as we will see in the next chapter, the liberalization
of international telecommunications ultimately generated sharp changes
and new instabilities.

Internet Takeoff

Through these means, a space was crafted in which a decentralized network
of networks could be sculpted. But the uses to which this space could be put
were not fully evident at any point; the idea of the Internet was not simply
waiting to be born. Nor was the eruption of the Internet during the 1990s a
function simply of the surging popularity of the World Wide Web and the
fast-paced widening of the email habit. Behind both of these lay the complex
rise of the personal computer industry during the 1980s and the subsequent
ascent of the local and wide-area networks that transformed standalone PCs
into a new communications medium.

These developments bring us back to business users. The proliferation of
PCs and the widespread competency training that went with them occurred
first on desktops in businesses.[63] Throughout the 1980s and 1990s, local area
networks (LANs) connecting PCs, peripherals, and other computing resources
likewise mushroomed mostly throughout major corporations. Urs von Burg

has shown how these electronic warrens were rapidly enlarged, in part owing to a canny vendor strategy of relying upon a nonproprietary, or "open," technical standard for their LAN products (Ethernet).[64]

Under the terms of successive FCC proceedings, still in line with the recommendations of Rostow's Task Force on Communications Policy, these LANs constituted "data-processing" networks, developing free of common-carrier regulation or public-utility status, by host companies and computer-industry vendors. By 1986, as intra-organizational networks and private value-added networks of different kinds burgeoned into a sprawling, deregulated domain, only two-thirds of U.S. network investment was made by public network carriers—down from nearly all as recently as 1980.[65] Expenditures by large corporate network users correspondingly rocketed.[66] But this was merely a preface to the network investment boom that crested at the turn of the millenium and transformed information technology in general and the Internet in particular into the core of the political economy.

The proliferation of corporate LANs helped spur an accelerating demand for cheap, wider-area interconnectivity. As Urs von Burg suggests, the Internet—employing another nonproprietary standard, TCP/IP—offered a prospective solution, and during the 1990s, corporate LANs "quickly became important on-ramps to it."[67] Perhaps two-thirds of the spectacularly increased Internet investment that occurred through the 1990s was undertaken by businesses, principally to erect walled-off private systems known as intranets,[68] while only one-third of such investment went to the enlargement of the public Internet.[69] As intranets were extended to establish new links between businesses, what began to be termed supply chains were rapidly reconstructed to take advantage of the new network capabilities. Not coincidentally, far and away the largest share of Internet e-commerce flowed between businesses, rather than between businesses and consumers.[70]

In the next chapter we will see that this great overhaul of telecommunications did not occur as an orderly growth process; nor did it institute an equitable, or even a smoothly functioning, new regime. Instead, the pan-corporate reconstruction of the telecommunications infrastructure engendered contradiction and crisis.

5

The Crisis in
Telecommunications

Into the second half of the twentieth century, socially inclusive electronic telecommunications infrastructures had yet to be established; for generations, telecommunications had tended to constitute a scarce business service. Haltingly and unevenly, during the decades after World War II, access finally began to broaden.

The United States, in the forefront of this welcome historical process, presided over unprecedented network extension between 1945 and the mid-1960s. Comprehensive residential access to the telephone was achieved after a delay of some seventy-five years after the telephone's invention by institutionalizing the expansionary policies crafted by New Deal regulators confronting a severe economic depression. Throughout war-crippled Western Europe and Japan, inclusive network buildouts took longer to accomplish, but by the 1960s and 1970s, welfare-state policies again were engendering widespread residential access. Across these developed market economies, network modernization programs incorporated important technological innovations. As electronic switching (circuit-allocation) systems and new transmission media were embedded in them, network infrastructures drew ever greater capital investments.

It was another story outside the developed market economies, where telecommunications infrastructures still languished. Throughout the global South, with modest exceptions, into the 1980s dramatically insufficient investment—a century-old scourge—continued to arrest development. Patchy domestic networks, scarred by gaping rural-urban disparities and further disabled by inadequate power supplies and technical support and the exploitive tactics of foreign equipment suppliers, yielded long subscriber wait-lists and unreliable telephone service. International connections came at a premium and were often vexatious. Were the supply of basic telecommunications to increase, throughout most of the world demand was certain to materialize.

During the 1980s and 1990s, the supply of global telecommunications underwent a new burst of expansion. The long-standing pattern of network investment likewise shifted. By 1997–99, perhaps half of world telecommunications investment was being absorbed by "developing and transition" countries.[1] Over a short interval, dozens of national wireline networks expanded, often incorporating up-to-date technologies. In Malaysia, the number of residential main telephone lines per hundred households grew between 1991 and 2000 from 36.7 to 64.5; in Honduras, from 7.9 to 16.2; in Argentina, from 28.5 to 68.7; in Botswana, from 5.9 to 24.6; in Cape Verde, from 11.8 to 53.7; and in China, from 1.2 to 33.9.[2] Wireless phone systems experienced even more extraordinary growth, as the world total rocketed from eleven million subscribers to a billion during the 1990s. A survey of Malaysia late in 2004 showed that nearly half the population used cell phones.[3] In a growing number of countries—nearly one hundred by the end of 2001—mobile phone users outnumbered wireline subscribers.[4] Scarcity began to recede; though dearth had by no means been vanquished, it was unquestionably diminishing.

Qualitative changes in technology—including, crucially, the Internet—contributed new service features to this process of network growth. One decade of commercial development found the Internet reaching most (88 percent) of the sixty million U.S. households with a PC by 2001.[5] Around the globe, hundreds of millions of people routinely use the Internet to send and receive email,[6] while very large numbers also employ the new medium to shop and window shop, listen to music, read text, and edit and exchange files coding photographic, musical, film, and video data.[7] Broadband (high-speed) Internet access has proliferated, engendering new applications and service packages from cable and telecom companies.[8] As system development proceeds, the Internet interacts across variegated sociocultural processes at every level, from the local to the global.

During the 1990s, this telecommunications buildup and the wider processes of change occurring around networks were heralded in euphoric terms. "Infinite bandwidth," sang a poet of U.S. high-technology stocks in 2000, would eliminate the crippling bondage imposed by "immobile information."[9] A benevolent wired world was arriving, a luminous republic of information predicated on newly abundant telecommunications.

These prognostications were summarily interrupted as it was recognized that in key markets a surfeit had developed. During the spring of 2000, a stock-market crash overtook the entire field of telecommunications, media, and technology companies. Over the next two-and-a-half years a couple of trillion dollars worth of capitalization in telecommunications was destroyed. In the United States, dozens of telecommunications companies went bankrupt. Lucent Technologies, the largest U.S. maker of telecom equipment, suffered thirteen straight unprofitable quarters through mid-2003 and recorded gigantic losses. Even the Internet plumbing manufacturer Cisco (though it did not cease being profitable) saw its stock price—which at the height of the boom had pumped up its capital to $500 billion[10]—plummet by 80 percent.[11] Layoffs throughout the industry—more than half a million as of August 2002[12]—eliminated more jobs than had been created during the preceding boom years (after 1996).

Some similar difficulties surfaced internationally. Bankrupt carriers like Viatel and hard-pressed telecom equipment manufacturers like Alcatel testified to the travails of the European market. As the capitalization of European telecom companies dropped by $700 billion between March 2000 and November 2002, the cumulative debt of the seven largest European carriers, one analyst observed in July of that year, was "greater than Belgium's gross domestic product."[13] Early in 2003 came reports of the largest losses in the annals of business of three major European countries: 9.5 billion Euros for the Dutch carrier KPN; 20.7 billion Euros ($22.7 billion) for France Telecom;[14] and $27 billion for Germany's Deutsche Telekom.[15] The debt loads borne by these carriers, meanwhile, had also grown frighteningly large. Across the globe, Japan's high-flying mobile carrier NTT DoCoMo saw its stock shed $180 billion in value between February 2000 and December 2002.[16] The debacle's ramifications, international as well as domestic, continue to be felt.[17]

Analysis of the telecommunications crisis—why it happened and what its chief results have been—is a vital task. Prevailing accounts are inadequate. The debacle did not originate in a group of corrupt stock analysts and errant industry executives, despite the publicity such individuals have garnered. Nor did it stem merely from the inflated projections of demand that motivated network operators and their backers to go ahead with infelicitous system-building plans. Rather, in vital ways the crisis was an outgrowth of the same institutional policy changes that had triggered the preceding network investment boom. The crash revealed that the market compulsions that had been blithely unleashed over the course of a generation by an extended community of corporate executives and investors, politicians, lobbyists, lawyers, and business journalists harbored fierce dangers.

*　*　*

Newly liberalized system-development policies were first forged in the United States. U.S. government authorities began, haltingly in the 1950s and 1960s and more systematically thereafter, to open segments of the long-restricted monopoly telecommunications market to entry by specialized service suppliers and network operators. Terminal equipment, value-added computer services, long-distance service, and finally, local service: at each piecemeal turn toward "pro-competitive" policy, an additional portion of the market was made available to new entrants. At each successive moment of liberalization, moreover, New Deal public-utility policies were further destabilized and eroded. Giant waves of capital generated by investors hungry for new outlets flowed into a long-closed industry, stimulating successive technological changes.

Aggressively foisted on the rest of the world by U.S. government agencies and multilateral institutions like the World Bank, this new regime was rapidly exported. During the late 1990s it attained global paramountcy. Market liberalization self-evidently served the interests of an enlarged group of telecommunications supplier companies and, not incidentally, their investment banks; but it also catered to the evolving demands of large corporate users, who repeatedly rebuilt their business processes and extended their markets around networks.[18] As a result, by the late 1990s, for the first time since the nineteenth century, worldwide telecommunications-system development was bolted to corporate-commercial foundations. Transnationally organized and operated networks functioned increasingly as big corporate capital's production base and control structure.

Huge outlays were needed for this. Through the 1990s, the financial markets answered nearly every call for capital by existing and would-be suppliers. As a bevy of entrepreneurs obtained cheap debt financing to build out new networks, existing giants such as AT&T and eventually newly privatized national network operators elsewhere joined the parade. Not solely in the developed market economies but throughout most of the world, telecommunications systems drew nearly unrivaled capital expenditures, amounting in 1999 to between 2.6 and 4.9 percent of each nation's total capital investment.[19] In many countries, the major carrier became the biggest single company on the national stock exchange.[20] Rival network operators, domestic and international, spent billions of dollars to build out systems with which to link office complexes throughout central cities. Corporate network users based in every economic sector expended additional billions on the tangle of hardware and software they needed to enlarge and modernize their burgeoning proprietary systems. By 1993, before the Internet came into widespread use, the business press was already showering praise on corporate America for having lavished a trillion dollars on information technology over the previous decade—and the real action, as we now know, was yet to come.[21] Much-publicized, network-related corporate investment functioned as the pivot of the late 1990s U.S. economic boom; its effects would be felt across the globe.[22]

Industry competition also intensified rapidly. Coming after years of policy liberalization, passage of the 1996 Telecommunications Act in the United States, followed the next year by the World Trade Organization Basic Telecommunications Agreement, together unleashed competition on a new scale. It became necessary for existing carriers to try to outflank their rivals using various combinations of financial maneuvers and investments in technological innovations to supply novel and/or more efficient services. Wireless providers, long-distance companies, local-exchange carriers, and—just over the horizon—cable television system operators, power-line companies, and burgeoning Internet-based telephone companies sought to carve out for themselves the largest share of the most profitable market segments.

Cross-border investment in what had been a field occupied by national flag carriers sharply accelerated. Vodafone, the British-based mobile-phone behemoth, used its high-flying stock to make nearly $300 billion of acquisitions— as a result of which by 2002 its affiliates claimed tens of millions of customers in twenty-nine countries.[23] Throughout Europe, recently privatized national network operators such as France Telecom and Deutsche Telekom mortgaged their futures to pay for pricey wireless franchises within their own national markets and to build up cross-border investments; at the height of the boom, during 1999–2000, France Telecom spent 88 billion Euros on acquisitions.[24] By comparison, the substantial outlays made by Japan's NTT DoCoMo on mobile phone companies outside its domestic market—$15 billion—seemed puny.[25]

The take-up of the Internet transformed this global scramble by subjecting it to a recurrent speculative phenomenon: market mania. The Internet, relying on qualitative changes in the technology of switching and vastly capacious transmission media, offered unmatched efficiency gains and could be used to support a rich array of information services. But in its volume and its effect on investors, self-serving puffery inflamed the Internet's development with a wildly excessive potential. In short order, the need for an "Internet strategy" became a corporate buzzword. Not only for carriers and equipment suppliers but for corporate users throughout virtually every industry, Internet investment at any price and for seemingly any purpose began to pass for strategic wisdom. In the most widely bandied estimate, which ultimately was shown to have been fabricated, "total Internet traffic" was "doubling every three or four months"[26]—meaning that, as far ahead as the eye could see, the bandwidth needed to transport the projected torrents of data would likewise have to be exponentially enlarged.

Wall Street bankers quickly calculated that such an atmosphere could be exploited to attract additional lucrative investments to infrastructure projects. Equipment spending by incumbent U.S. local telephone companies doubled to around $100 billion annually between 1996 and 2000, even as technological breakthroughs qualitatively boosted information-carrying capacity and efficiency on long-distance systems. Network operators continued to throng into the market; initial public offerings of their stock made many of their investors and executives fabulously rich. Celebrity executives broadcast via serried ranks of industry analysts, PR staffers, and cheerleading journalists that their state-of-the-art systems would supply ever more voluminous quantities of bandwidth at ever lower prices. Each new entrant testified to the successful promotion of the need for additional capital spending. And each rendered what was already a competitive environment increasingly cutthroat.

A final, seemingly remote macroeconomic trend also played an important part. In response to the Asian financial crisis of 1997–98, the United States opened a flood of easy money. The result was to subject the already super-heated stock market—and the information-network sector that had become its poster child—to the full force of capitalism's "animal spirits." Years had been spent preparing that sector for investors, and now dollars, yen, marks, pounds, francs, and other currencies poured in. A speculative tidal wave was forming. Outrageous hype now passed for sagacity. Projections of exponential increases in demand for capacity functioned as the only common coin needed to entice further investments.

Much-ballyhooed capacity-building innovations added fuel to the fire. One particularly significant innovation was dense wave division multiplexing (DWDM), which allows signals transmitted at many wavelengths to be carried by existing fiber-optic networks; exalted by George Gilder and many others, DWDM hugely expanded the capacity of existing network circuitry. Harnessed to the institutions of speculative finance and the policies of market liberalization, technological innovation became a destabilizing force that went far beyond anything imagined, or desired, by self-appointed proponents of Joseph Schumpeter's concept of creative destruction. Because of the sudden immense expansion in the supply of rivalrous network systems, demand—though it never ceased to increase—simply could not keep up. Investment became overinvestment, surplus became surfeit, competition became destructive; but these results, even after they became undeniable, were generally presented as a function only of mistaken demand estimates and crooked executives.

Well before the crash it could be seen that the foundations of the system were shaky. At the margins, a scattering of analysts began to worry that particular market segments might be in danger of becoming overbuilt. Andrew Odlyzko, a former AT&T researcher who had moved to the University of Minnesota, publicly challenged the inflated forecasts of Internet demand. In a 1998 report for the Economic Policy Institute on WorldCom's proposed takeover of MCI, I forecast that, on financial and other grounds, this gigantic merger was "a mistake waiting to happen."[27] In 1999, I went on to "raise the question of whether global telecommunications . . . might be headed toward a market glut, that is, a state of secular overcapacity."[28] But the "vortex" that I imagined, when it materialized, went beyond anything I could have forecast.

At roughly this point (1999–2000), to placate investors and shareholders, executives at several U.S. companies began to cook the books. WorldCom, which eventually fell into bankruptcy, was found to have claimed no less than $11 billion in spurious revenues and was forced to pay a record $750 million fine for its malfeasance. Top management at Qwest, another important carrier, engaged in what was ultimately labeled pervasive fraud by claiming $3.8 billion in nonexistent revenue, according to a complaint filed in 2004 by the Securities and Exchange Commission; Qwest responded by agreeing without imputation of guilt to pay a $250 million penalty.[29] Sprint's aggressive use of tax shelters for accounting purposes forced its two veteran executives to quit; they were soon rehired to head a Japanese carrier that had passed through a leveraged buyout to the New York investment company Ripplewood Holdings.[30] These were all multibillion-dollar carriers sitting near the top of the industry's food chain. Insider trading, shady financial practices, accounting fraud, ties between commercial and investment banking services, and other forms of corporate corruption ran wide and deep.

But the underlying problem—to which many of these tactics comprised a response—was overcapacity within a context of market competition. Suddenly, after a generation of networking the market system as a corporate-commercial project, the profound vulnerabilities introduced by that same project began painfully to be expressed. These, unfortunately, were not limited to a few miscreant firms but ramified across the industry's length and breadth. Soon enough they would threaten not only the telecommunications system but the larger economy.

Systematic appraisal of the extent of overcapacity is problematic, but there were estimates of 35 percent and, in some cases, of 2 percent utilization rates for long-distance circuits. And that condition of overcapacity has persisted. In December 2004, analysts reported that 80 to 90 percent of the U.S. fiber-optic circuits installed during the telecommunications boom still sat unused; six months later, researchers estimated that "less than 5 percent of the total transmission capacity of all the fiber lines is being put to use—about the same amount as in 2001."[31]

The shape and extent of this overhang were raggedly uneven. Qualitative enlargement of information-carrying capacity occurred principally along the most profitable, high-density traffic routes. Between and within the world's metropoles and along major transoceanic routes, new systems added unprecedented increments to available network capacity. Between 1990 and 2004, for example, according to a knowledgeable forecast in 2001, "[N]ew submarine cables will have multiplied trans-Pacific bandwidth on the order of 15,000 times."[32] In chapter 9 we will examine some of the ramifications of this Pacific surplus for China and the world economy. But it is not unique. One new trans-Atlantic fiber-optic system planned to make available more capacity than had been provided before by all existing trans-Atlantic networks.[33] Between 1998 and 2003, submarine and terrestrial circuits in service between the United States and other countries leapt up by 778 percent.[34]

Residential access to telecommunications, however, was a different story. Even in the United States, where 95 percent of households possess telephones, home-based Internet access remains scanty among minorities and the poor, as compared with whites and higher wage-earners[35]—yet the Bush administration has sought to eliminate programs to subsidize Internet access for schools and libraries, where PCs were accessible on a public basis.[36] New network infrastructures often still do not even reach poor rural districts throughout Latin America, South Asia, and Africa. "As always," a candid industry analyst observed in 2001, "economically less-developed regions and countries may have a long wait."[37] Continuing inequalities thus suffuse and sharply limit the modernization of telecommunications systems. As Pippa Norris observed,

> In 2000, most of the world's online community (84 percent) lives in highly developed nations. In comparison, some thirty-five societies classified by the UNDP with low levels of human development, such as Nigeria, Bangladesh,

and Uganda, contained about 1 percent of the online population, although home to half a billion people.[38]

In 1999, an American needed to save a month's salary to purchase a PC; a Bangladeshi had to save all his or her wages for eight years to do so.[39] The rapid emergence of new services around the Internet have thus led to what one study called "social exclusion on a global scale."[40] Still exhibiting telephone access levels common to the United States before 1920, the countries of Latin America bore widespread witness to what one analyst in 2005 called "Digital Poverty."[41]

Throughout Africa, especially, residential telephone access continues to be desperately inadequate (in some countries, perhaps one to four phones per hundred people); but what is to be expected when even a capital city like Conakry enjoys electricity only one day in four? In numerous countries, moreover, charges for consumer phone service are becoming *more* inequitable. Long-distance rates have declined, but this mainly benefits business users and middle-class residential callers. Local rates, in contrast, remain persistently high. In Indonesia, with forty million unemployed out of a population of 212 million late in 2002, sharp increases in telephone rates mandated by the International Monetary Fund were only rescinded in January 2003, after tens of thousands of workers and students took to the streets in protest.[42] From Brazil and Argentina[43] to South Africa[44] and Southeast Asia, the global crisis in telecommunications has halted the enlargement of access that had commenced during the 1990s; millions of subscribers have actually dropped off national networks. And, again provoking protests, as a result of deliberate U.S. policies, international telecommunications revenues flow increasingly out of the countries of the global South and thus cease to be available for national development.

Even where telecommunications exist in surplus, however, the effects of previous policies are now punishing. The press focuses with fawning concern on one affected party: the industry's corporate participants. No segment of the telecommunications industry has emerged unscathed.[45] Giant equipment manufacturers such as Lucent and Nortel, some of whom have pumped up their own sales by helping to finance unknown startup carriers, are now suffering calamitous revenue declines. Lucent Technologies, which had loaned $7.5 billion to its customers and lost $29 billion between 2001 and 2003, cut its payroll from 157,000 in 2000 to 32,000 in 2004 and has sold, closed, or spun off twenty-seven of the forty businesses it had purchased since 1996.[46] Lucent's seeming return to profitability in 2004–5 actually reflected not improved sales of equipment but income accruing to its pension fund. During the go-go years, Nortel had undertaken a $30 billion string of acquisitions, leaving it with a market capitalization of $270 billion by mid-2000. Now it became enmired in accounting scandals and was compelled continually to restate its financial results for prior years—while it fired thousands of employees.[47] Nortel took charges in nineteen of twenty-four quarters from 1998 through 2003, losing $34 billion between 2000 and 2002.[48] Specialized market entrants, which mushroomed overnight in the heady days of the tech boom, have also met difficulties. Corning, which diversified from consumer products into fiber-optics to profit from the boom, found its new business precipitously wrecked. In 2000, Corning's fiber-manufacturing revenues accounted for 74 percent of its $7.1 billion in revenue overall; by 2003, operating only one plant and having closed or abandoned four others, fiber manufacturing made up 40 percent of total revenue of $2.7 billion.[49] JDS Uniphase, a company hailed by George Gilder in 2000 as "a potential Intel of the telecosm" for its role in fashioning optical components, had cut its work force, which had reached thirty thousand, to eight thousand by late 2002.[50]

Transoceanic fiber-optic cable networks—some only recently completed,[51] others still under construction—now chase buyers. Global Crossing, a $20–billion Bermuda-based company whose capacious, state-of-the-art submarine network had been allotted much more publicity than the equally newsworthy composition and political connections of its corporate board or its lavish campaign donations, which helped it gain access to top levels of the U.S. government, declared bankruptcy in 2002 amid allegations of chicanery.[52] Dozens of new providers of local services in the United States summarily shut down. These so-called competitive local-exchange carriers had invested tens of billions of dollars to enter the market, in no small part because regulators, enamored of the siren-song of "competition," forced their rivals, the much larger incumbent Bell companies, to lease them network facilities at a deep discount.[53] By one estimate, even as the competitive local-exchange carriers succeeded in opening the business market to competition, their total market capitalization (inclusive of publicly traded carriers only) dropped 96 percent between 1999 and 2001, from $86.5 billion to $4 billion.[54]

However, the social damage that has resulted from the crisis exceeds the business travails borne by individual carriers and the financial woes faced by industry executives and investors.

Buried by the publicity accorded to a succession of business success stories during the boom, large superfluous costs and other irrationalities and abuses engendered by decades of liberalization have now taken their toll.

Years before the crisis erupted, the industry's embrace of competition had rendered it ever more dependent on advertising. Each competitor also has had to build up a separate and mainly duplicative management and staff. Regulators find no reason to object to these or to the regulatory costs that come with competition, as authorities struggle to cope with the new regime. Within the dominant policy-making discourse, these costs continue to lie submerged.

The chairman of the Federal Communications Commission, Michael Powell, had to eat humble pie when he conceded in February 2003 that "few of the Commission's actions in implementing the Telecommunications Act of 1996 have produced identifiable benefits to the American public."[55] But this admission still gives no hint of the fact that, virtually without exception, federal regulators of both major parties had clamorously supported policy liberalization, and that they continue to do so. Nor did Powell's speech point to a closely related trend: that lobbying and campaign-finance expenditures by the mutually antagonistic industry segments and individual carriers launched by telecommunications liberalization have helped keep the whole shell game in motion.

Liberalization exacts its highest price from the most vulnerable populations. Prison inmates in the United States have been turned into a literally captive market, as carriers sign lucrative contracts with underfunded municipalities to kick back a percentage of the phone charges they extract from the growing ranks of the incarcerated and their typically poor families; by the 1990s, the prison telephone industry had grown into a billion-dollar business.[56] The effects of the liberalization process on telecommunications workers—again, inadequately covered by the press—constitute a more basic ravaging feature.

Newly created jobs at startup competitor companies are increasingly shorn of any right to collective bargaining; the number of union workers in telephone and data services dropped by over half between 1985 and 2004, from 625,000 to fewer than 275,000. According to a different estimate, the proportion of telecommunications-industry employees represented by unions in 2004 was 24 percent.[57] Pay and benefits show the effects. Millions of additional call-center employees across the length and breadth of industry labor for low pay in high-tech sweatshops. As international networks make such moves more cost-efficient, call-center jobs are sent in growing numbers offshore to India, the Philippines, and the Caribbean. With continuing layoffs by carriers, the quality of residential telephone service naturally suffers, while local rates for residents, as opposed to business users, remain stubbornly high.[58] It was cold comfort to hear one genuinely concerned FCC commissioner declaim at a lowpoint that "we must establish a concrete plan for how we will protect consumers in the event a carrier ceases operations or otherwise disrupts service."[59]

The meltdown furnishes a grim lesson in information-age economics, as problems within the networking sector rapidly ramified. After a decade awash in funds, suddenly there was virtually no new capital available. Investors grew every bit as reluctant as they had earlier been rash in doling out funds, because they now anticipated that asset values would only continue to decline. Existing carriers cut back sharply on network buildouts and upgrading projects. Of course, there continued to be demand: people and organizations still needed communications, the crash notwithstanding, and on a historically impressive scale. But the crisis created unyielding rigidities, aggravated further by the narrow focus on quarterly profits that had come to typify the liberalized, Wall Street–oriented industry. Pricing pressure therefore remained intense, especially in crucial corporate and wholesale service markets, where a continuing price war led to repeated reports of prices set below cost,[60] while, until 2004, equipment sales continued to stagnate.[61] All this contributed to leading carriers' increasingly beleaguered condition.

Overcapacity inflamed by free-market policies, technological innovation, and a juggernaut of speculation now led to cannibalism. Local carriers diversified in part to devour their own existing markets, as subscribers to their wireless subsidiaries substitute wireless for wireline service and email for voice telephony, and as they offer high-speed digital subscriber-line services that eliminate residential customers' need for second phone links. The number of U.S. telephone lines served by existing local carriers actually declined during 2001–2 for the first time since the Great Depression; by mid-2004, the Bell companies that dominated local service had lost twenty-eight million lines, a drop of 18 percent since the end of 2000.[62] The carnage has been worse in long-distance markets. Whereas in 1995, the average AT&T long-distance subscriber made 143 minutes of calls per month, by 2004 that figure had declined to a little over sixty minutes—owing again to customers' increasing use of wireless phones, email, and text-messaging.[63] In wireless markets throughout the developed market economies, prices began to decline as market saturation developed; in the United States, as wireless carriers competed for customers, the average per-minute cost of a cell-phone call declined by more than 65 percent over the four-year period to early 2005.[64] Weakened by price wars and continuing product substitution, the top three U.S. wireline long-distance carriers (AT&T, MCI, and Sprint) posted multibillion-dollar losses, fired thousands of employees, and took huge charges against earnings to write down the value of assets.[65] Even this has proven insufficient to stanch the industry's wounds. As its bonds were reclassified as junk and as regulators threw their weight to the large, politically connected local-exchange carriers (mainly the four Bell companies), the long-distance carrier AT&T—for a century the controlling hub of the U.S. telephone system—announced a historic withdrawal from the residential service market.[66]

The core of the older public-utility telecommunications system is being hollowed out. Selectively bypassed by major corporate users, the existing system is being attacked for different reasons by big local-exchange carriers, cable system operators, Internet service providers, and other purveyors of new services, from voice-over-Internet companies like Vonage to makers of video-game consoles possessing Internet capabilities. Traditional forms of rate regulation and long-standing strictures of common carriage have come under fierce pressure from carriers and new industry participants demanding to supplant them with privately negotiated contracts for pricing and service.[67]

As the industry's room for maneuver contracted, what had originated as a problem afflicting the networking sector triggered more general economic worries. In one estimate, more than a third of 2002's record $157 billion worth of worldwide corporate debt defaults came from distressed telecom companies. Actually, at this point the sector's total global debt load was far greater: at a minimum, between $500 billion and $1 trillion.[68] Major banks—J. P. Morgan Chase, Citigroup, Bank of America, Deutsche Bank, ABN Amro, and Toronto Dominion—held huge quantities of this debt, and it bulked large in some of their own financial losses. In June 2002, European Union banks held telecommunications, media, and technology debt amounting to 17 percent of their own funds; though their exposure thereafter trended downward, the European Central Bank remained sufficiently worried in February 2003 to warn that careful monitoring was still required.[69] How did these blue-chip banks come to be so highly leveraged? During the halcyon days of the boom, their investment and commercial units fell over themselves in their scramble to underwrite stock offerings and to provide loans to the carriers.

Individual banks took measures to shield themselves. They were greatly aided by the novel speculative instruments they and their rivals created as a result of the interrelated process of financial deregulation; many depend in different ways on deregulated computer communications. Enormous quantities of telecommunications debt were repackaged and sold off; through bewildering new financial products like collateralized debt obligations, credit-default swaps, and derivatives, and through well-established practices like bond sales and syndicated loans, banks hived off tens, perhaps hundreds of billions of dollars' worth of debt onto insurance companies, pension funds, hedge funds, and other banks. In dispersing this debt, individual institutions reduced their own risk *by increasing systemic risk.* It was hardly reassuring that no one appeared to care exactly how these vast obligations had been reallocated; "The risk is out there somewhere," blandly conceded one financier early in 2003.[70]

Despite these measures, the crisis continued to intensify during late 2002 and early 2003. In underlining its severity in his address to a Goldman, Sachs investment conference, the FCC chairman Michael Powell employed extraordinary language: "The status quo is certain death and can no longer be considered a viable option."[71] A few months later, when the billionaire investor Warren Buffett suggested that, because their effects are so little understood, derivatives—much used by telecommunications-industry lenders—constitute "financial weapons of mass destruction," his remark called forth a quick, soothing dissent from no less a figure than the Federal Reserve Bank chairman Alan Greenspan.[72]

And then, suddenly, the urgency eased, and the crisis was over—or so it appeared. High-visibility executives were hauled before the cameras by law enforcement officials and charged with fraud and other crimes of corporate looting. While their cases inched through the legal system, newly appointed managers and special bankruptcy-court-appointed officials sorted through the debris.[73] Restructuring continued to be accomplished, however, through ad hoc business moves taken in response to self-interested and sometimes fractious investors seeking payouts rather than through a socially inclusive process of system development. Executives require ratification by regulators for large-scale changes, but in the dessicated public sphere of the post-9/11 period, meaningful debate has been virtually absent.

In another era, it is easy to imagine the crisis engendering quite another response. Let us indulge the fantasy that a "truth and reconciliation commission" has been formed to inquire into the origins and character of the debacle. Such a body would need to gather testimony not only from convicted and unconvicted telecommunications executives but also from an array of other complicit actors: major investment banks and industry analysts; accountancies; top law firms; the legislative, judicial, and executive branches of government; think tanks; regulatory agencies; business users; academic economists; and corporate-commercial news media. Of course, our hypothetical commission would also have to grant a public hearing to those who were victimized: employees, pension-fund contributors, retirees, and residential users. The commission could thus focus a wide-ranging public debate about the structural first principles and underlying social purposes of telecommunications. Debate would subject to careful scrutiny the liberalized system's chief policy premises: that networks should be built out and reconfigured to cater to the specialized needs of transnational business users and to serve the financial interests of the investing class, with only secondary regard for the interests of workaday employees or ordinary residential users.

The telecommunications crisis, like the overall liberalization process that preceded and produced it, then could be seen for what it is: not merely an industrial and financial debacle but a crisis in social and political function. Truly to address it, in an attempt to rebuild the sector on a sound structural basis, would require policy makers to abandon the shibboleth of free-market competition as a sufficient corrective to monopoly. Carriers typically leagued together with corporate users: Ford Motor contracted with SBC to design, build, and manage a $100–million Internet phone system for fifty thousand employees at 110 locations.[74] By this means, they contrived together to bypass and further subvert the existing regulated network system. "Pro-competitive" policies merely sought to ensure that Ford had a choice of suppliers in accomplishing this purpose.

If, however, the proper regulatory goal were not a competitive industry structure, what should it be? Power to direct and shape system development in accord with societal needs. By directly confronting the growing primacy of corporate capital—carriers, equipment suppliers, and business users—over other social interests, regulators could restore the accountability of the political economy's networking function.

The political traction needed to establish such a radically different regulatory trajectory—or to initiate even a modest public debate—remains absent. Yet the idea is not merely fanciful. A so-called Telephone Investigation was mandated by Congress and mounted by the FCC in 1935 in response to an earlier crisis in U.S. telecommunications. That investigation gave rise to the most extensive and critical public portrait of the industry ever produced and generated sharp and enduring changes in policy for U.S. telecommunications-system development. The contrast with this prior precedent could hardly be clearer. Once more, organized public debate is virtually absent; nothing like the Telephone Investigation has even been mooted.

Contemplating a different policy trajectory, one framed in an earlier generation's terms for thinking about representative democracy, would have to commence with a comprehensive audit of the industry's financial condition. Yet, tellingly, liberalization has rendered such a review problematic: the very data on which regulators base policy are increasingly self-reported by carriers, proprietary secrets held by business users, or supplied by the same Wall Street analysts who helped throw the industry into crisis. The FCC commissioner Michael Copps may be applauded merely for venting his frustration at this state of affairs: "We must commit to doing the hard work of collecting our own data rather than relying on potentially misleading and harmful financial, accounting, and market information produced by corporate sources subject to clear biases and market pressures."[75]

In its apparent aftermath, recognition that the crisis was caused not by a few miscreants but by a more basic political-economic process has rapidly dwindled. It has not been effaced entirely within mainstream discourse, but where it does surface, it is carefully canalized. The *New York Times* all but pinned the rap on WorldCom CEO Ebbers's felonies: not only had the industry leader, MCI, been tainted in its takeover by WorldCom, declared the *Times*, but WorldCom's "phantom growth caused once-mighty telecommunications companies like AT&T to cut prices and slash costs in the crippling race to keep up, from which they never fully recovered."[76] All true enough; but these are symptoms—not basic causes—of the industry's prolonged travails.

A complementary approach has also developed. A journalistic exposé of the "broadbandits" whose plundering brought the industry to its knees and whose depredations involved collaboration by regulatory, financial, political, and media institutions, concludes: "[T]he telecom bubble that burst was part of a painful but necessary cycle. . . . [T]he broadband bubble will soon become a distant memory."[77] By naming the crisis as a normal outcome of market-based development, its wider ramifications are sidestepped and contained. Market forces might have demonstrated their dark side, but as a result progress once more will commence: no pain, no gain.

This view expresses a basic ideological concordance with the winners. The economist Hal R. Varian demonstrated the familiar logic of this alignment early in the gathering crisis: "Overinvestment may seem wasteful, but then it's always easy to identify winners once the race is over. . . . One of the great strengths of American-style capitalism is its ability to finance crazy ideas—because every now and then, those ideas have a very, very big payoff."[78] The question of who obtains this payoff and who gets stuck with the bill in the form of what mainstream economists call "externalities" (lost jobs, a democratic deficit in policy making, and so on) is thereby evaded.

Far from offering an ideological safe harbor, the role of market forces in the debacle might be made a matter for debate. This has not occurred. Even as the industry again turns to lobbyists, regulators, and politicians to stem its travails,[79] it has been conveniently forgotten that the crisis itself stemmed from a profound failure of institutional policy, and that network-system development remains lodged in the same institutional hands and serves the same institutional interests that it did during the runup to the debacle.

For this reason, it may be said that, far from having ended, the crisis in telecommunications has now merely entered a different phase. From the beginning of U.S. policy liberalization—and this constitutes the desideratum of the entire, piecemeal initiative—proponents have demanded that corporate capital be afforded the same ability to deploy and control telecommunications networks as it typically assumes over other forces of production. Policy makers acceded to this demand, throwing open the telecommunications market to hothouse innovation and expansion by competing units of capital. The financial meltdown of the industry shows that the system-development process that ensues from this radical policy shift has been made to incarnate the contradictory market logic of the larger political economy: *more private control* for corporate users, network suppliers, and investors translates into *less societal control and reduced democratic accountability.*

No sign has been given that changes have been worked into this foundational policy doctrine. Responses by state agencies and major corporations, purportedly redressive, have mainly comprised efforts to carry on the same process of proprietary corporate reconstruction around networks. Rather than engaging the essential social and political features of the crisis, they have tried in selective and partial ways to ward off or contain some of the destabilizing forces that liberalized system development has generated. It is too early to judge how this new round of regulatory interventions might further reshape the telecommunications system. The primary intentions motivating this response, however, can already be glimpsed.

Government and business strategists are working to reduce the overhang in network capacity or, at least, to induce capital once again to look with favor on the industry. Significantly, both major presidential candidates during the election of 2004 stressed a need for ubiquitous broadband (high-speed Internet) service deployment.[80] To aid in rationalizing this overbuilt industry, political support cohered for a new round of previously unthinkable mergers by big carriers, as SBC acquired AT&T and Verizon took over MCI. The motive that fired the two top-dog carriers to try to carry off these colossal takeovers was merely a self-interested hope of altering in their own favor the terms of trade with business users.[81] The *Wall Street Journal* admitted as much in an editorial: "[B]oth AT&T and MCI have been exiting the residential market and concentrating on big corporate customers, which is what these acquisitions are really about."[82]

Two other facets of this response stand out. Executive, legislative, and regulatory agencies cooperate with private companies in tightening administrative oversight of what is now dubbed a "critical infrastructure." Ostensibly to enhance the nation's coordinative capacities in the "War on Terror," this fusion of government and industry attests to an attempt to exert top-down control—with a strengthened potential for repression and civil-liberties abuses.[83] Meanwhile, in a succession of true emergencies, the September 11 attacks, the 2003 blackout, and Hurricane Katrina, the neoliberal network infrastructure performed abysmally.[84] Regulators also attack so-called asymmetric regulation for being an antiquated and hindersome legacy because it mandates one set of rules for cable TV companies, another for telecommunications carriers, and a third for spectrum-dependent broadcasters. Using the supposed new reality of "convergence" as a rationale for jettisoning "the silo approach that we've had since essentially 1934,"[85] regulators have sought instead to allow communications companies of every kind maximum freedom from accountability in their renewed attempts to develop multifunctional proprietary systems, notably around Internet and mobile services. Simultaneously, political pressure is ratcheted up to squelch or reorient efforts to provide such systems by municipalities rather than corporate-commercial interests.[86]

These measures have deepened the antidemocratic tendency that has marked liberalization from the beginning. On the one hand, remote bankruptcy courts now seek to adjudicate the contending demands voiced by different classes of investors, while scanting those of the general public. On the other hand, under the guise of homeland security, specialized agencies have removed governmental telecommunications decision making from already inadequate oversight mechanisms. Crucial policy powers thus devolve on ever more obscured executive-branch agencies like the Committee on Foreign Investment in the United States and the Critical Infrastructure Protection Board.

Secretive deliberations and backroom deals will do little to enhance the security of the telecommunications system, at least if security has to do with the workaday needs of ordinary people; nor will rate hikes for household consumers, layoffs for employees, and continued network development on behalf of investors and transnational business users and military agencies. But why, we should ask, do regulators not only support further mergers but also, paradoxically, press for policies supporting continued competitive market entry, renewed network building, and additional rivalrous investment? While bankrupt carriers such as MCI and Global Crossing are permitted to reenter the market, shorn of debt and ready to compete again in a context of declining price levels, new carriers and cheaper forms of provision—most importantly, voice-over-Internet services—also find ready authorization.[87] The way has been cleared, meanwhile, for head-to-head competition between cable system operators and local-exchange carriers.[88] Does all this not make for a glaring policy anomaly—a mistake? Is it possible that regulators have not absorbed the lesson of the crisis: that unbridled network building and competition were responsible for the industry's devastation? Why try it again? Is the authorization of escalated competition, even if prospectively ruinous, a stark reflex of the continuing hold exercised over policy-making zealots by free-market ideology?

In part, yes, if we add that the skids of ideology are still well-greased with campaign donations, revolving-door corporate jobs for erstwhile regulators, and lobbying expenditures. But complementary structural realities loom even larger in this renewed liberalization attempt. Although a general financial crisis has been averted, the telecommunications industry as it existed has been irretrievably destabilized. A new system must be erected in its stead. Ultimately, the drive to build this new system in accord with corporate capital's aims and priorities persists because network systems and services are still, and are still seen to be, primary axes of whatever systemic growth there may be. Though the process of system development may inflict additional casualties, profitable growth for big capital overall remains superordinate as a strategic matter.

Although George Gilder's vaunted "telecosm," with its promise of "infinite bandwidth," has turned out to be a carriers' graveyard rather than an investors' cornucopia, on the demand side business users in many markets and application areas have enjoyed unprecedentedly cheap access to a widening field of network systems and service innovations. If supply-side instabilities can be alleviated by authorizing previously unthinkable mergers between carriers, good; but even if they cannot, liberalized policies ensure that the economy-wide process of corporate reconstruction around networks will remain free to develop as its proprietary overseers prefer. Corporate-commercial priorities rather than public-service obligations will continue to predominate.

In the international context it has become especially evident that policy makers are willing to risk significantly destabilizing entrenched processes of system development. National network operators that, for one reason or

another, had escaped or waited out the storm, by the early 2000s found themselves suddenly positioned to exploit exceptional opportunities for transnational expansion—mostly at the expense of ailing U.S. carriers and their
backers. To refocus its strategic attention on the roiling changes in its home
market, for example, BellSouth sold off the extensive wireless networks it had
established in South and Central America. Spain's Telefonica purchased most
of its systems, adding them to its 38–billion-Euro-investment in the region,
to become by 2004 the largest regional wireless operator, with forty million
subscribers spread through twelve countries, and the third-largest carrier
worldwide.[89] More striking still was the emergence of Telefonos de Mexico,
controlled by the billionaire Carlos Slim, as a transnational carrier. Telmex
had preserved its home-market dominance in Mexico despite fierce competitive pressure from units of AT&T and MCI. During 2004, however, as both
U.S. carriers staggered, Telmex snapped up distressed wireline-network assets
from MCI in Brazil, and—at around one-tenth of their initial cost—from
AT&T in Argentina, Brazil, Chile, Columbia, and Peru.[90] In another case, a
consortium led by China Netcom, China's second-largest carrier, paid $80
million to acquire a regional affiliate of the bankrupt Global Crossing, carrying a book value of $1.2 billion; soon after, China Netcom bought out its
partners to become the sole owner.[91] Videsh Sanchar Nigam Ltd., a unit of
India's Tata Group, purchased a thirty-seven-thousand-mile network linking
three continents from a U.S. conglomerate; the seller, Tyco International, had
already written off its entire disastrous investment in the system and sold it
for $130 million—a discount of $3.5 billion from its initial cost.[92]

The passage of ownership and control of these properties to capitalists
based in the global South represents a historic sea change. As units of capital based in Mexico, China, and India acquire supranational networks at
rock-bottom prices, previously marginal, sometimes nonexistent, corporate
interests have suddenly found means of joining what had been an exclusive
international club, akin to the possession of nuclear arms. Social movements
beyond capital are also able to selectively draw on surplus telecommunications capacity. We will see in chapter 5 that projects that took advantage of
that surplus but were not attributable to it—like the establishment of Al-
Jazeera by Qatar—signify that the crucial field of international television
news has been destabilized.

Two mutually antagonistic trends have contributed to this dynamic, contradictory context. On the one hand, entry into international telecommunications by Chinese, Indian, and Mexican capitals—not to speak of already
substantial Japanese and European corporate interests—is certain, at least

in the short term, to confer added strategic leverage on prospective rival powers; existing intercapitalist rivalries are therefore likely to intensify and to engender increased multipolarity in the continuing process of capitalist political-economic construction. This, in turn, is likely to occasion heightened ideological and policy divergences as well as policy conflicts over the swiftly changing global division of labor, as outsourced work continues to be pushed and pulled into low-wage countries.

On the other hand, the continuing strategic depth and sweep of U.S. corporate and state interests in telecommunications should not be underestimated. As U.S. leaders are well aware, the newfound abundance of supranational network systems, built around Internet technology, likely will render mere possession of physical telecommunications circuits less important for control of international communications than it has been traditionally. The power to configure and reconfigure routers, the Internet's controlling instrumentation; to shape and alter the Internet's logical attributes through the domain-name system; and to develop new supranational service offerings and surveillance and management controls remains in every case highly skewed toward U.S. interests. Internet governance has thus bid fair to become an increasingly vital pivot of policy and, over the longer term, of international and intercapitalist conflicts as well.

The crisis in telecommunications occurred as part of an even more encompassing transformation, as accelerated commodification gripped communications, culture, and information. To carry forward the analysis, we turn next to consider the culture industry.

6

The Culture Industry: Convergence and Transnationalization

A hallmark of modernity has been the sustained expansion of a technologically dynamic corporate apparatus for the manufacturing and sale of culture. Within the developed market economies, by the early twentieth century, "art with commercial intent," as one approving executive calls it,[1] had begun to permeate industrial design, and trademark-based brand advertising was aimed at magazines, newspapers, and other cultural commodities.[2] Media companies' efforts to sell directly to consumers on a national scale likewise accelerated. Around older cultural commodities such as books and newer ones such as musical recordings and films, they innovated the techniques of the blockbuster hit: star personalities, distinctive genres, mass publicity and marketing tie-ins, and wide distribution. Extruding internationally over the decades, this corporate-commercial culture industry grew so encompassing that, as Eric Hobsbawm comments, it "drenched everyday life in private as well as in public with art. Never had it been harder to avoid aesthetic experience. The 'work of art' was lost in the flow of words, of sounds, of images, in the universal environment of what would once have been called art."[3]

The growth of the culture industry has registered profoundly on the experience of social collectivities and the consciousness of individuals; its effects have been, however, historically mutable, complex, and sometimes contradictory. In this chapter, however, I will focus not on these effects but on the political-economic processes that generate these ceaseless flows of cultural commodities. By tracing the contours of the trend toward accelerated commodification into the culture industry, we may see not only how the terrain on which capital logic operates again has been enlarged but also how the forms and effects of this capital logic themselves have been altered.

These changes were again predicated on struggles to neutralize or deflect a variety of obstructions, pressures, and constraints. Three types of historically interrelated obstacles have inhibited capital in its attempts to accelerate the culture industry's growth: political barriers and protests, intra-industry conflicts over system development, and technical and territorial limits on existing markets for cultural commodities.

Over the last quarter-century, substantial ground has been gained on all three fronts. Contradictions and impediments—some of them newly emergent—persist, and market reorganization remains volatile and perpetually unfinished. However, elites have employed corporate and governmental power to overcome political adversaries and prior market constraints. Through these means they have gone far toward reconstituting cultural production and distribution on a more fully capitalist basis, a changed technical foundation, and a transnational scale. The establishment of a global capital logic for one of informationalizing capitalism's leading industries has been no mean feat. How did this momentous transformation originate? How has it been shaped and channeled? What are the leading facets of change?

Convergence

Beginning in the 1970s, a succession of new media began to appear, often supplementing or modifying existing forms of experience, but sometimes offering more novel departures: VCRs and DVD players, CD and MP3 players, game consoles, mobile telephones and other wireless devices, PCs and notebook computers. At the same time, existing distribution networks, built around terrestrial broadcasting and telephone infrastructures, began to be joined by new transmission media—satellites, optical fibers, and wireless systems—while each of these distribution systems enlarged its role as a prospective carrier of voices, images, and data rather than, as had been generally true, only one of these modes. Tremendous techno-economic dynamism powered a sustained expansionary surge, some of the contradictory effects of which have been surveyed in previous chapters, across a vast field of network systems and applications, semiconductors, and consumer electronics.

Again propelled by the United States, this combined trend toward multimedia and multifunctional networks was often seen in terms of an overarching process of communications-industry "convergence." In 1983, Ithiel de Sola Pool offered what has become a paradigmatic explanation:

A process called the "convergence of modes" is blurring the lines between media. . . . A single physical means—be it wires, cables, or airwaves—may carry services that in the past were provided in separate ways. Conversely, a service that was provided in the past by any one medium—be it broadcasting, the press, or telephony—can now be provided in several different physical ways. So the one-to-one relationship that used to exist between a medium and its use is eroding. . . . The explanation for the current convergence between historically separated modes of communication lies in the ability of digital electronics. . . . [E]lectronic technology is bringing all modes of communications into one grand system.[4]

Twenty years later, the idea that, as a congressional committee heralded in 2004, "convergence is blurring the lines between voice, video, and data services" remained a touchstone in policy-making discourse.[5]

The idea of convergence emphasizes the importance of the technological upheaval in communications and validly stresses that communications infrastructures are becoming multifunctional as they assimilate versatile digital electronics. Prophets of convergence insisted, however, that this transformation occurs through an intrinsic technological imperative; Pool concedes that his argument amounts to "soft technological determinism."[6] I will argue, to the contrary, that convergence stems from a trio of complex changes linking science and technology development, industry strategy, and public policy.

Claims for technological convergence in communications actually originated not in the 1980s, but half a century before. They emerged after the ability to theorize scientifically about, and alongside the ability to engineer, networks that could carry anything that could be given the form of an electronic signal—voices, texts, and images. Efforts accordingly multiplied to expand the capacities of electronic networks to accommodate diverse signals. As David A. Mindell has shown, this trend took hold well before the creation of the first electronic digital computer.[7]

"The common bases of electrical communication," a leading engineer and communications-industry executive declared in 1935, had been gradually discerned over the preceding sixty years.[8] Frank Baldwin Jewett singled out a pair of dramatic manifestations from 1915, when AT&T engineers instituted transcontinental telephony by wire and experimented on transoceanic telephony by radio.[9]

Throughout the 1920s and 1930s, AT&T researchers amassed detailed knowledge of electronic communications, encompassing telephony and telegraphy, radio broadcasting, talking motion pictures, sound recording, and television. This was possible because they recognized that these seemingly disparate areas actually possess a shared identity—they comprise what Jewett, the head of Bell Laboratories, referred to as "variant parts of a common applied science." "One and all," Jewett elaborated, "they depend for their functioning and utility on the transmission to a distance of some form of electrical energy whose proper manipulation makes possible substantially instantaneous transfer of intelligence.... Further, whenever and wherever we can transmit electrical energy to a distant point for one form of communication, we can, in general, use that same energy for all other forms of intelligence transmission." Media that rely on electrical energy constitute, he underscored, "integral parts of a single common whole."[10]

This insight allowed an equally sweeping perspective on future system development. One of Jewett's engineering colleagues declared that "[l]ooking to the future, the writer likes to think of radio and wire telephony as increasingly dovetailing together to form one general front of advance. The principles and technical tools being fundamentally the same for both, a technical advance in one is likely to help the other.... As regards service, we observe that radio and wire telephony are one and the same thing in respect to the over-all result, that of the delivery of sound messages."[11] Jewett agreed that communications-system development would increasingly pivot, in practical terms, on this underlying identity. "From a scientific and engineering standpoint it is apparent that our future course will result in structures and systems by which every form of electrical communication can be given over a plant which is largely interchangeable and readily adaptable to changing service requirements."[12]

Much remained to be accomplished, however, before integrated system development could advance. During and after World War II, new research began to point the way. Incorporating theoretical contributions by Claude Shannon (also of Bell Labs) and others, digital computers, telephone switches (circuit-allocating mechanisms), and other electronic systems began to be engineered to treat "information" as a "common currency ... carrying generalized signals and messages through the continuous world as discrete pulses."[13]

For this engineering change to become the basis for system development, a reciprocal shift in business organization was required. Technological convergence unfolded only as it merged with a second process of change: a long-standing effort by communications companies to extend their businesses across existing technological limits, industry boundaries, and program forms—seeking vertical integration, economies of scope, risk reduction, and higher profitability.[14]

From its outset, the telephone business was intermittently intertwined with the telegraph. Throughout the twentieth century, telephone executives sought repeatedly to extend into what they saw as organically related media businesses, arguing that the industry as it existed was irrationally and opportunistically divided. Jewett stressed the "somewhat fortuitous way in which the several fields of electrical communication have developed" and credited this—alongside errors of thought and "the vested interests involved . . . which think of themselves in conflict"—for having made it impossible to unify the industry under the aegis of the telephone company. "Possibly our worst sins in this direction are to be found in the fields of business and governmental supervision and control," Jewett hinted, in an admission that seems artfully candid as well as studiously vague: "To some extent these wholly artificial and misleading classifications have been and are consciously perpetuated by those whose special interests will be served thereby."[15]

Already in Jewett's day, communications executives were targeting the fragmented industry and the discrepant laws and regulations that had grown up around its different segments as hindrances to the pursuit of profitable accumulation. In this predigital era, attempts were being made to bring under one roof various discrete systems and services via common corporate ownership. That some of these systems had been unified by physical theory was, in this context, merely an argument to support such ambitions. Industrial convergence in communications comprised a manifestation of a much wider organizational tendency within corporate capitalism.

Industrial convergence was driven by the omnipresent corporate need to maintain or increase profitability in a context marked by technological dynamism and the boom-and-bust cycle. During the 1920s and early 1930s, for example, newspaper owners acquired ownership stakes in numerous radio stations to defend their turf—the nation's news markets—and to gain a new source of advertising revenue, while RCA, the nation's top radio manufacturer and broadcaster, extended its business by taking over the Victor record company and investing in the Hollywood film studio RKO. Strikingly, however, through the first three-quarters of the twentieth century, industrial convergence in communications remained sharply limited, arguments about common scientific principles notwithstanding. This was because the prevailing policy—the third foundational dimension of convergence—tended to support rival companies and, to a lesser extent, employee and consumer interests in repeated clashes over dominant corporations' bids to cross over into adjacent markets.

Far from being granted a mandate by policy makers to enter or establish new industries, AT&T—the giant in U.S. communications—was consistently prevented from doing so.[16] In 1913, AT&T was forced by the U.S. Department of Justice to divest its controlling ownership interest in the telegraph giant Western Union. Locking horns with General Electric and Westinghouse during the mid-1920s over the right to control development of radio broadcasting, AT&T again was pressured by government agencies to back away not only from broadcasting but also from other new media—sound film and musical recording—in which it held a decided interest. Even though AT&T's national infrastructure continued to carry the broadcast networks' signals, the carrier was unable to control the broadcasting business proper. Moreover, 1934, the year before Jewett's public frustration over "misleading classifications" and "government supervision and control," saw passage over AT&T's objections of a federal communications act that formally divided "broadcasting" from "common carrier" telecommunications, subjecting each domain to separate rules and regulatory policies. Finally, in 1956, the Justice Department took legal action to push AT&T out of the business market for computers.

Though the record shows a pattern of prospective AT&T involvement in virtually every major subfield of modern communications, with minor exceptions, the company was repeatedly denied the ability to participate extensively in market-development processes outside voice telephony.[17] More generally, as Pool emphasized, albeit with considerable unevenness, bursts of federal rule making erected prohibitions or limits on cross-media ownership by communications companies based in different industry segments and relying mostly on separate distribution infrastructures. Into the late 1960s, regulators instead mandated a complicated and far-reaching series of industry divisions. Regulatory walls limited computers and telecommunications, as they had earlier separated broadcasting from common-carrier transmission. Cable-television system development was undertaken under yet another rule-making mandate. The print media, finally, were also substantially differentiated, in legal and policy terms.

Discrete industry segments—common carriers far more than broadcasters or cable television companies—were, furthermore, subjected to public accountability obligations. A contradictory mix of factors provoked these patchy forms of institutional oversight, which were imposed during the 1930s and 1940s: the state's overarching effort to interdict the economic crisis of the Great Depression; corporate lobbying; and political opposition to corporate power stemming from trade unions, consumer groups, and New Deal regulators themselves. Regulators also harbored disparate objectives: to stimulate economic growth; to stabilize the industry within discrete oligopolistic market segments, typically by frowning upon market entry by new competitors; and to restrain and qualify unbridled corporate control over communications.

Common-carrier oversight was by far the most stringent, though there were large—and lucrative—loopholes. To major communications companies, nevertheless, existing regulations eventually came to constitute real obstructions to profitable growth. By the later 1960s, again for disparate reasons, they recommenced their efforts on behalf of industrial and technological convergence. The notoriously high profitability enjoyed by commercial telecasters, at least throughout major urban markets, excited interest in market entry via new modes of distribution such as cable systems and satellites. Growing saturation of communications-equipment markets, slowing growth rates, prompted manufacturers to cast about for new sites of profitable investment. Facing intensifying pressure to dispose of the commodities they produced because of the growing global glut, consumer-products companies experimented with the development of new media for advertising.

Perhaps the most powerful pressure originated, however, from a seemingly remote source. Across the length and breadth of the U.S. economy, industrial users and specialized corporate-equipment suppliers were seeking to implement new or revamped applications of computer communications throughout workaday business processes. To gain license to deploy data communications, they lobbied for fundamental policy changes in the regulated telecommunications system. Absent such policy changes, their ability to innovate around computer communications would be progressively hindered. From within and without the existing culture industry, therefore, came mounting demands to relax and even to withdraw legally codified market divisions and restrictions on technological-system use and market entry. The unsettled relations between business and consumer markets for converging network services continued to present difficult policy challenges.[18]

Rapid engineering advances in digitization were repeatedly incarnated in product innovations by computer, telecommunications, and consumer electronics companies. Digitized products and services in turn presented a continually widening array of technical opportunities or entry points at which attempts to realize new forms of industrial convergence could commence. Matching the level of abstraction sought by Jewett fifty years before, a growing portion of human experience now could be objectified and represented, with varying verisimilitude, as digital signals: "the esperanto of 1s and 0s," as two analysts called the process.[19] Digital signal coding, processing, and transmission drew momentum from a series of equally general economic advantages: capaciousness, flexibility, and accuracy in processes of signal reproduction, storage, and transmission. Digital systems also were recognized for their substantial efficiency, allowing greater reliance on machine perception and handling of information.[20]

But these remained abstract potentialities until technological convergence within functioning digital networks and services could be actualized. Policy liberalization was the sole means by which this could be made to occur. Only through policy changes could space be created in which to deploy digital technologies to build novel and prospectively profitable pathways between once-separate areas of practice. Relaxing existing policy strictures was necessary even for incremental steps to be taken toward multifunctional services provision by regulated corporate proprietors of what had been discrete infrastructures. Liberalized policies were again needed before media owners could seek to merge diverse programming units, service capabilities, and distribution systems according to executive whims, consultancy fads, or what proved to be shrewd strategic vision. The path toward convergence was not, therefore, a mere reflex of imperatives resident within today's communication technology. Before it began to be employed rhetorically as a cause, convergence functioned as an objective—even an agenda—for policy change.

Unevenly, beginning around 1970 and gaining strength thereafter, arguments for industrial and technological convergence attained sufficient political traction to attack market-segmenting policies. Through episodic and specifically bounded policy shifts, convergence began, finally, to attain limited practical expression. This change must be understood within the context of a much wider reversal of official priorities.

An important part of the drive to renew profitable accumulation for U.S. corporate capital proceeded by attacking old and new government supports for economic, social, and political welfare. Rarely adequate, sometimes dominative in practice, these programs nevertheless attested to at least a modest commitment by the state to collective bargaining rights, a minimum wage, unemployment insurance, social security benefits, environmental protection, occupational safety, and equal employment opportunity. To the unremitting frustration of the political Right, the communications and culture industry had graced this welfarist political order with its ideological cover—the unstated assumptions and templates of argument that granted it public legitimacy. And so, at this juncture, "convergence" served not only as a means to overcome barriers to profitable market growth in communications but also as an entry point for a pan-corporate assault on prevailing policy priorities.

Three specific initiatives proved most vital. First, officials began to relax rules limiting further concentration of ownership and cross-media owner-ship in communications; thus they created an important carrot with which to tempt existing owners. Second, they created a stick by beginning to target the market divisions and service restrictions that had separated the different modes of telecasting, telecommunications, and print media. Prohibitions on ownership and service linkages between these industry segments were epi-sodically withdrawn. Each of these shifts was consolidated over time; with the AT&T divestiture, passage of the Telecommunications Act of 1996, and the eruption of the Internet, the liberalized policy space became increasingly definitive.

These twin changes in structure and practice prompted a third initiative: a campaign to secure the new policies by developing a supportive ideological framework within which they could appear to make sense. This effort turned on discrediting the limited existing mechanism of public-service account-ability. During the initial period of piecemeal policy change, Pool began this project by identifying convergence as a danger to freedom of speech. A "shadow" was darkening, he argued, as the fetters of state intervention, long evident in broadcasting and telephony, now threatened to extinguish vener-able liberties as they were generalized in the emerging era of digital electron-ics. Convergence presented policy makers with a choice as worrisome as it was inescapable: either definitively extend the tentacles of government control, or accede to the potentialities of the vaunted "technologies of freedom" on which Pool placed his hopes. For Pool, the legal traditions adhering to the print media function democratically and could be counterposed to all the tainted forms of government regulation that had been overlaid onto broad-casting and telecommunications. During the years following publication of his landmark book, however, the ideology of "convergence" evolved further. The purported evils of regulatory intervention continued to be assailed, but other issues associated with the technologies of a supposed informational abundance were assigned increasing emphasis.

Presenting convergence to the public as an inexorable technological and economic necessity, policy makers have insisted that their own supreme function is to speed development of competing multipurpose distribution systems, or "platforms." "Advancing technology," declared the president's Council of Economic Advisors in 2005, "is providing competitive inroads to a number of industries once considered natural monopolies." As satellites converge on cable-television service providers and wireless telecommuni-cations converge on wireline systems, "technology-induced competition" negates "the natural monopoly rationale for the economic regulation."[21]

Well coached by trade associations, public-relations agencies, corporate technologists, mainstream economic consultants, and industry lawyers, the politicians and bureaucrats who produce policy as a formal matter have equated the "public interest" in communications more and more rigidly with the supposedly natural and desirable state of marketplace competition. In Section 401 of the 1996 Telecommunications Act, competition is enshrined explicitly within this industry segment as the desideratum of the public interest.[22] Contrived thereby was a clever, though spurious, justification for marginalizing social, political, and economic concerns that have periodically erupted into communications-policy discourse—notably, the foundational issue that concentration in communications constitutes a class-based threat to democracy. Convergence was to engender greater marketplace efficiency, which in turn would obviate any further need for public accountability.

The role that remained for regulators, for rhetorical purposes, anyway, was to undo what they had done before—often, ironically, on behalf of direct ancestors of leading culture conglomerates. Free-marketeers complained, as formerly distinct industries and companies integrated and new rivals searched "for ways to offer consumers a bundled set of communications, broadband, and even media services under a single brand name," that prevailing policies continued to mandate that "increasingly interchangeable services . . . be regulated under different legal standards."[23] Lawyers and academic economists made much of "asymmetric regulation" and the growing need for "regulatory parity": no individual communications-industry segment, they now held, should have to abide by regulations that are not also imposed on other, now converging, segments.

Proponents' claims on behalf of regulatory parity seem simple: as convergence deepens, policies have to be altered to ensure that all service providers operate within a uniform regulatory environment. We should not ask more of a telephone company than of a cable system owner, as each scrambles to provide consumers with a comparable basket of video, voice, and data services. To think otherwise is to abet discrimination. "A regulatory regime that still treats substitutable platforms differently will distort the marketplace by, among other things, creating artificial regulatory advantages for one set of competitors over another."[24] As this exhortation implies (by foregrounding marketplace efficiency), official policy deems one of the two obvious policy options for inducing regulatory parity altogether impermissible.

I refer to the idea of reconstituting an enlarged and strengthened set of public-service responsibilities and planting them across the length and breadth of the converging industry. Merely transposing traditional U.S. common-carrier tenets of nondiscrimination would only begin to make a start on this complex and difficult project, not least because the world—rather than solely the nation—is the relevant jurisdiction. Nor would policies framed for print, or cable, or broadcast media resources prove sufficient. The very foundations of the public-service principle need to be reimagined and thence reapplied.[25]

Even under more benign circumstances, the success of such an initiative would require radical intellectual effort and great collective political will. The prime procedural requirement is simply stated: We need to organize workable but far-ranging practices of democratic authority over the global system of cultural provision. Substantive measures must include strong limits on commercially funded culture together with enlarged public funding; universal access to systems and services; robust data protection, with guarantees of freedom from corporate and government surveillance; and much-enlarged rights to define and freely use common cultural resources in different formats. Each of these requirements carries beyond prior precedent.

By this emboldened standard of public responsibility the quality of actually existing policy practice appears not merely insipid but persistently abusive. In place of a full and fair debate to devise policies that meet social needs and take meaningful advantage of network potentialities, there are only endless echoes of industry's self-serving demand to be freed of the dead hand of government oversight. Similar attempts to discredit public accountability have acquired commanding influence outside the United States, while in the United States, the historical center of the convergence process, free-marketeers continue to repeat their wearisome refrain:

> In a world where consumers have multiple wireline and wireless options, Congress must recognize that the old rationales for regulation have been either satisfied or rendered moot by the relentless march of technology.... [T]he first step Congress must take to begin seriously reforming communications policy is to end this asymmetry, not by "regulating up" to put everyone on an equal footing, but rather by "deregulating down." Placing everyone on the same *deregulated* level playing field should be at the heart of telecommunications policy.[26]

The dominant approach treats with contempt the idea that networks should be shaped, overseen, and used with regard for any substantial public interest. The contempt has been concealed, however, by a veneer of theory that, by equating the public interest with marketplace efficiency, purports to vitiate the former. To take the measure of this theory it is enough to recall again that, for all the hallowed chatter about economics and efficiency, communications-industry convergence—like the economy-wide transformation of which it forms a part—is actually predicated on repeated political interventions on behalf of one special interest: business. No one should doubt that planning and coordination are essential features of contemporary communications systems, as they are of big business in general. Regulation, often derided and reviled, for this reason will continue. The real question is, who will undertake the planning, and what social purposes does regulation serve?

In the United States over the course of the past generation, the answer has been consistent. Liberalization was freed to occur as a matter of private accumulation strategies rather than public service. Not surprisingly, as liberalization has proceeded—first in the U.S. domestic market, then internationally—capital has flooded into the culture industry, and the size and scope of profit-taking endeavors within the sector have expanded convulsively. Convergence has entered a new phase. Through an endless and ever-changing array of corporate accumulation strategies, digitization of networks and network applications has extended. As turf battles continue to be fought between rival groups of corporate suppliers and between equipment providers and users, the communications system is unevenly reorganized around long-heralded common foundations.

Bulking Up: Conglomerated Culture

In engaging with this complex and dynamic process of system development, existing communications- and culture-industry owners in the United States—the world's largest domestic market—have been compelled to grapple with two strategic business issues. Audiences are migrating away from conventional radio, television, newspapers, and magazines, thereby eroding the mass-market, or "broad-reach," media model that once predominated. In the United States, average hours per person per year spent with broadcast television declined from 867 in 1999 to about 778 in 2003; with daily newspapers, from 183 to 173; and with consumer magazines, from 134 to 123. Meanwhile, average hours per person per year spent with cable and satellite television increased from 720 to 949; with video games, from 53 to 75; and with the Internet, from 80 to 169.[27] Unless owners can wrest control of these and other new media, their major advertisers' access to coveted audiences will erode. Owners also recognize, however, that their continued market dominance—and perhaps their very existence—requires that they seek jurisdiction over programming assets and communications outlets.

Otherwise, corporate rivals may inflict damaging new competitive pressures. Even in advance of capacious digital-network buildouts, therefore, communications and culture conglomerates began to coalesce, as rivals sought to control overlapping service arrays and distribution systems.

Beginning in the mid-1980s, what has proven to be a continuing round of multibillion-dollar mergers and acquisitions commenced. In 1982, the largest U.S. media company was the American Broadcasting Company, claiming revenue of $2.7 billion; by 1998—a couple of years *before* its spectacular $100 billion coupling with AOL, in what would be one of the largest mergers in history—Time Warner already constituted the biggest U.S. media company, with revenue of $28 billion. During 1991, U.S. media mergers amounted to $14 billion. In contrast, in the first three months of 1999—near the height of the boom in technology and Internet stocks—$300 billion worth of U.S.-based media, entertainment, and telecommunications deals were announced.[28] The shock-effects of the burst financial bubble that followed carried the reorganization of the communications industry into a different phase. Now, corporate indebtedness rather than the gravity-defying movement of share prices, vulnerability rather than strength, triggered further industrial reorganization. However, the cascade of mergers and acquisitions only recommenced.[29] In 2003, *after* a plunge in its market value and a huge charge against earnings, Time Warner—still the market leader—was a $40 billion behemoth.[30] Though strategies have continually shifted, and assets have been accordingly shuffled and reshuffled, a handful of companies have emerged to dominate the process of reorganizing culture and communications.

Qualitative organizational changes were the predicate of the quantitative shift in scale. Increasingly, each conglomerate began to develop strategies to multiply the functions performed by its distribution channel or channels—cable systems, terrestrial broadcast systems, satellites, and wireless and wireline telecommunications networks. Internet technology has increasingly acted as the means by which to move baskets of voice, video, and data services down each of these transmission systems. Again with the support of regulators, investment banks, and cheerleading journalists, the conglomerates are successfully appealing to convergence to broaden their asset bases and to take in the most profitable program-production properties, encompassing everything from film and television programmers, print publishers, recorded music companies, professional sports teams, game publishers, theme parks, and Web services.

As one index of the rapid corporate integration of programming and distribution infrastructures, during the decade from 1992–93 to 2002–3, liberalized rules allowed the four major U.S. terrestrial television networks—each now a tentacle of a larger conglomerate—to increase their overall ownership share of the prime-time programs they aired from 32 to 76 percent.[31] In such circumstances, not surprisingly, parent conglomerates were well equipped to assimilate emergent areas of commodified cultural production, such as video and computer games, that had been spearheaded and cultivated by independent capital. Diversified giants like Time Warner, Sony, News Corporation, Disney, NBC-Universal, and Viacom tracked consumers across once-discrete media frontiers, sold them a range of services directly, and offered major advertisers access to an array of media with which to stage the sales effort. By 2005, NBC's mainstream television business accounted for less than 15 percent of group earnings as a subsidiary of General Electric; early in the 1990s, NBC earnings had represented more than half the total.[32]

Efforts to identify "business models" through which the culture industry might pursue profits drained prodigious quantities of energy from business schools, the financial press, and consultancies. Stripped down to the basics, there are two familiar possibilities and two emerging potential alternatives. I will illustrate them by reference to the Web. First, direct sales of content could be made to organizations and/or individuals via subscriptions, site-licenses, one-time payments, or rentals. Although reliable data are scarce, U.S. consumer spending on online content, exclusive of gambling and pornography, by one estimate increased 14 percent in 2004 to $1.8 billion.[33] Sales of content to organizations were assuredly higher. Second, advertising could be sold to third parties so that commercial corporate sponsors could be said to "support" such forms of cultural practice. During 2004, U.S. online advertising expenditures increased to around $10 billion. The third and fourth business models are strongly linked with online systems: transaction charges imposed for using a given service or application, and data sales, where information about users, audiences, and customers is cumulated, categorized, packaged, and sold to interested third parties—often, again, corporate advertisers. Across the length and breadth of the culture industry, companies that can count on multiple sources of revenue have begun to build up a new competitive advantage—for example, in financing blockbuster program extravaganzas.[34] The largest conglomerates have come to depend on this increasingly variegated economic base not only to help amortize the rising costs of programming but also to experiment with the market development of new cultural commodities.

The still-unfinished unification of the industry has left a profound impress on the language of the cultural marketplace, which has become symptomatically abstract. Not long ago, executives were still prone to speak of books and films and recordings. During the 1990s, these usages began to seem quaint; the increasingly preferred terms of reference are "content" and "product." This change bespeaks a strategic mission that has transformed as convergence has proceeded: to repackage, or "repurpose," different kinds of cultural material across different media has become an overarching corporate goal. This purified accumulation strategy is deliberately—strategically—indifferent to genre and text as repositories of discrete form and meaning, often hostile to the "moral rights" of authors, and dismissive of any larger concern for cultural inheritance. It expresses the universalizing market ambition that has always suffused the corporate drive for convergence.

A popular film thus routinely generates a television series, a musical recording, a video game, and/or a novelization, to say nothing of the deluge of licensed merchandise that travels in its wake. Major publishers of trade and mass-market books "conceive of books as only another derivative product" and aggressively experiment with new electronic media.[35] In March 2000, a milestone was passed when Viacom's Simon and Schuster distributed four hundred thousand copies of Stephen King's sixty-six-page novella *Riding the Bullet* over the Internet in a single day. Taking note, Bertelsmann's Random House subsidiary, the world's largest book publisher, announced that it would digitize nearly its entire list (including 3,500 current titles and more than twenty thousand active backlist titles) to sell print-on-demand titles over the Web.[36] Though the venture proved to be premature, the intent to distribute print works electronically has mushroomed dramatically during the mid-2000s. By 2005, the question had shifted to which companies—publishers, or new online intermediaries like Google, Yahoo, and Amazon—would control this new market.

Books, however, no longer functioned only as discrete products. Random House had already established a division, RH Entertainment, "to create children's properties that can be licensed for development as books, toys, films, or in other formats."[37] As one U.S. journalist declared, "The smartest people in the business understand a basic truth: Books are probably the cheapest way to test an idea or a story. It costs maybe $75,000 to publish a book. You can't make a TV series show for less than several million dollars. Think of book publishing in this way: as an efficient form of market testing for ideas and for stories."[38]

The consumer-marketing complex, which propelled a tidal wave of commercialism across new and old media alike, is never far from the center of the new strategic thinking. An illustrative example may be drawn, again, from publishing. Makers of candy bars, breakfast cereals, and baked goods, including such companies as Mars, General Mills, Sun-Maid Raisins, Pepperidge Farm, and Nabisco, introduced "educational books" as marketing media, either as promotions or full-price for-sale items. Two breakfast cereal (Cheerios) books sold more than a million copies—enough to qualify them as best-sellers—in one year following their publication in 1999.[39] Films and video games are also routinely transformed into new venues for commercial product placement.

Alongside control of conventional and new electronic distribution channels, monopoly privileges (copyrights) undergird this cultural production-and-distribution system. Possessed of legal rights to control cultural properties spread across a lengthening list of media, the big culture and communications companies have built up their holdings of creative works; the world's largest music publisher, EMI, possesses rights to more than one million songs.[40] Conglomerates are acutely aware of their strategic dependence on this property. During the mid-1990s, for example, the president of Viacom declared: "'Viacom is fundamentally a software and copyright-driven company.'"[41] Aiming to capitalize on DVD sales, in 2004 Sony paid nearly $5 billion to acquire Metro-Goldwyn-Mayer, thereby establishing a joint holding of eight thousand movie titles.[42] Enlargement and aggressive enforcement of these legally constituted corporate privileges comprises a leading feature of informationalized capitalism.[43] The culture industry has been in the forefront of this police action.

The corporate abstraction of "content" and of "intellectual property" has dovetailed with the newly versatile technical basis that the culture industry is creating through its political embrace of convergence. During the 1990s, the unquestioned locus of digital convergence became the Internet, which was used to add transactional features to the distribution of an increasingly comprehensive range of cultural commodities. Before sketching key features of this still-emerging system, however, we must return to the fact that the market-development pressures that have been unleashed quickly broke through the confines of the U.S. domestic political economy. By the 1980s, it was evident that the very emergence of a conglomerate culture industry betokened the invention "of means to expand the market output to a global scale."[44]

Transnationalization

At the beginning of the postwar era, Western European countries—the largest markets for U.S. film exports—had renewed earlier attempts to protect national cultural space by imposing restrictions on the deluge of Hollywood exports. Governments forged these policies for diverse reasons and with considerable variation, but with only modest success, as Hollywood outflanked them by turning to coproductions and foreign direct investments.[45] Major trading partners then attempted to deflect the powerful outward projection of U.S.-dominated commercial television. As of 1959, Great Britain, which had accepted commercial television in 1954, limited imported shows to 14 percent of the total television schedule, while Canada employed a looser 45 percent quota. Japanese authorities deployed foreign exchange restrictions, which mandated a maximum payment of three hundred dollars per thirty-minute imported episode—far below the going market price.[46] In the changed context of the 1990s, dozens of countries continued to try to protect national culture—or, at least, domestically based audiovisual industries—via coproduction schemes involving subsidies, favorable tax treatment, import quotas, and other measures.[47]

Less-remarked-upon but more substantial barriers to accumulation also persisted in the implicit and explicit limits placed by many countries over advertising and profit-seeking investment in cultural domains. Into the 1980s, culture was typically subjected to norms of nation building and/or public-service responsibility, however inadequately formulated and incompletely actualized.[48] Directly and indirectly, great swaths of cultural activity were concurrently supported by government or nonprofit funding. Throughout much of the world, legally formalized property rights in cultural commodities remained all but nonexistent. Electronic channels of diffusion were limited in geographic range and lashed to single-purpose distribution systems organized within and by nation-states. Capital in general, and transnational capital in particular, therefore had only begun to acquire direct control of cultural production and distribution; and such control as it enjoyed was neither comprehensive nor stable.

Of greatest urgency by the 1970s, a political movement for a New International Information Order (NIIO) was gaining strength. In the United Nations framework of multilateral agencies such as UNESCO, with contributions from writers, artists, journalists, academics, and a variety of organizations, struggles on behalf of democratic cultural restructuring came to comprise perhaps the foremost barrier to accelerated commodification. Dozens of Asian and African states that had just freed themselves from European imperialism joined "dependent" Latin American countries in targeting cultural domination as a primary and persistent obstacle. Independent development priorities, these countries insisted, should preempt claims over domestic audiovisual space by foreign (principally U.S.) purveyors of cultural commodities.

As the 1970s waned and the 1980s dawned, U.S. government agencies and their international institutional affiliates took aggressive countermeasures to undercut the NIIO movement. The Thatcher government and the Reagan administration withdrew their respective countries from UNESCO; the United States, which had comprised the largest national contributor to the organization, did not rejoin for nineteen years (until 2003). This move signaled that genuinely pluralistic, multilateral discussion of global culture and information policy was at an end: "a warning to the international community that leading Western powers would not be outvoted by the majority of the world's nations," in the apt summary of Kaarle Nordenstreng.[49]

Around the same time, forceful bilateral pressure on behalf of augmented corporate freedom to invest in and use networks began to be exerted by the U.S. Trade Representative and the Department of Commerce, as well as by private trade associations and individual companies. The same impetus achieved a multilateral expression in the General Agreement on Tariffs and Trade and its successor, the World Trade Organization. Through structural adjustment programs overseen by the World Bank and the International Monetary Fund, dozens of countries privatized their telecommunications systems in whole or in part. Pressure to develop new outlets for a world already awash in surplus capital flooded the networking sector with investment.

Privatization was accompanied by the construction of a wide variety of supplementary network infrastructures, as a result of which proprietary systems became available for transnational conglomerates to distribute cultural commodities. Often self-servingly, U.S. authorities and network equipment vendors offered donations and limited funding for corporate-commercial system buildouts. And the International Telecommunication Union, reshaped to confer on private corporations rights that had previously been possessed solely by nation-state members, began to "'advise countries to dismantle structural regulations preventing cross-ownership among broadcasters, cable operators, and telecom companies.'" In this top-down synchrony, more than 150 countries introduced new telecommunications legislation or modified existing regulation during the decade beginning in 1990, and "convergence" steamrolled across the world.[50]

Additional research is needed, engaging diverse national contexts, to develop a more comprehensive analysis of the profound changes that followed. Already, however, some of the basic patterns may be glimpsed.

Internationally, through the 1950s, television remained mostly an enclave service for the affluent and, as one contemporary writer put it, basically "a

Western monopoly, with a Japanese extension."[51] But the new medium was growing unevenly but quickly. By 1965, 176 million TV sets were in use worldwide. As Wilson Dizard wrote, "In a decade it has spread from a few hundred transmitters grouped largely in the North Atlantic area to over four thousand stations on five continents."[52] Thirty years later, there were 1.16 billion sets in use, and television had moved definitively beyond the rich states into virtually all national cultures.[53] TV enjoyed an especially substantial surge during the 1990s. Between 1989 and 1998, the number of sets in Brazil leaped from 30 million to 53.8 million; in Mexico, from 11 million to 25 million; in South Korea, from 8.8 million to 16.1 million; in Indonesia, from 9.4 million to 28 million; in India, from 22.5 million to 70 million; in Iran, from 3.6 million to 10.3 million; in Morocco, from 2.3 million to 4.5 million; and in South Africa, from 3.5 million to 5.5 million.[54] In 1996, a milestone of sorts was passed when more television sets were in use in less-developed countries (692 million) than in developed nations (669 million).[55] China, with over 360 million sets in 1998—up from merely one million twenty years earlier[56]—now boasted by far the world's largest domestic prime-time viewing audience.[57]

This continuing impressive enlargement of TV equipment sales—the reception component of the global audiovisual infrastructure—was complemented by new openings for market development throughout the system's production and distribution segments. After an extended interval of slower development, and despite continuing inequalities in provision that were most extreme throughout sub-Saharan Africa, the culture industry began to undergo an increasingly multifaceted and frenetic reorganization. World trade in cultural commodities, inclusive of cinema and photography, radio and television, printed matter, literature, music, and the visual arts, quadrupled from $95 billion in 1980 to more than $380 billion in 1998.[58] Over this same span, while global trade in general underwent a continuing sharp increase, the share of cultural commodities in world imports rose from 2.5 to 2.8 percent.[59] This growth was exceptional in a context of widespread economic stagnation and crisis, especially throughout the countries of the global South. In Latin America during the "lost decade" of the 1980s, the communications industry was nearly unique in enjoying robust growth.[60]

For decades it had seemed that the international commercial expansion of the new medium and the overarching American stewardship of this growth were indivisible.[61] As a sector previously hedged with restrictions on capitalist exploitation was substantially opened to it, many new entry points for commodification were established by, or in association with, U.S. interests. In Latin America during the late 1980s and the 1990s, in addition to new terrestrial broadcast stations, pan-regional cable and satellite networks owned by U.S. conglomerates like Time Warner, Viacom, News Corporation, and Disney were established. Not only did these transnationals succeed in expanding their subscriber bases; they also helped trigger a swift consolidation of multichannel service provision within different national markets.[62] An entire institutional complex was refashioned, from cable systems and program networks to audience measurement services and more. As John Lawrence Sullivan observes, "[T]he largest buyers of advertising time on U.S.-based cable networks are U.S. advertising agencies representing large multi-national firms (many based in the U.S.) for a wide array of consumer products such as soft drinks, running shoes, and cars."[63] Programming and advertising were often coordinated out of Miami, Florida, which, Ana Fiol suggests, began to serve as "the media capital of Latin America."[64]

National capital, however, has also often claimed a significant role in implanting and reorganizing this enlarged transnational complex. Sometimes freshly constituted, national capital has often sought and gained substantial access to its home market and, on a more selective basis, even expanded into foreign markets. At the capillary level, but notably accommodating some movement by domestic capital into the ranks of the "second-tier" transnational companies,[65] a scattering of conglomerates domiciled in countries such as South Korea, China, India, Mexico, Brazil, and Venezuela have thereby joined those based in the United States and a few other wealthy nations in widening and deepening the culture market—and extending the entire industry's transnational orientation. New media services and freshly minted cultural commodities play a strategic part in opening venues to these commercial actors. Computer game exports, to take an important example, increased fourfold in value over twenty years (1980–98), and a leading role in purveying console and game sales was wrested by companies based in Japan, South Korea, and China.[66] Like an array of other economic sectors, the culture industry is being rearticulated to accomodate greater market participation by non-U.S. companies based in many countries.

Interweaving with this crucial general change are other trends. To begin with, the process of reorganization is decidedly *not* open on equal terms to all units of capital. According to Richard Parsons, the chief executive of Time Warner, the "genius" of his rival, the CEO of News Corporation, Rupert Murdoch, stems from his ability to "see the whole world as the game board."[67] A handful of transnational conglomerates is uniquely equipped to strategize and act on this same panoptic basis. Together with dominant corporations spread over a wide series of newly adjacent fields—a nonexhaustive tally includes semiconductors (Intel), network infrastructure (Cisco), consumer software (Microsoft), and wireless telecommunications (Nokia)—they comprise the spearheads and chief beneficiaries of the entire process, as Edward S. Herman and Robert W. McChesney have emphasized.[68]

A second-tier of perhaps a hundred national and regional companies also plays an important role in the transnationalization process. National capital's importance is as much political as economic: under the sign of neoliberal reconstruction, it typically allies with the leading transnationals behind a politics stressing convergence and industrial consolidation. The terms on which the bounty is to be shared constituted a source of endless bickering, and, as John Sinclair has underlined, even large regional and national players at points find their room for maneuver limited by actors "higher up on the food chain."[69] Nevertheless, on the foundational principle that culture should be a profit-making business, there has emerged a striking consensus among a significantly widened global class of entrepreneurs, investors, and executives of every stripe. As Fiol puts this in the context of Latin America,

> [T]he mainstream media (press, radio, and television) took an active part in this process of structural reform. They acted to speak up for and to legitimize the good of changing the economic model and its new rules of the game. At the same time, they were direct beneficiaries of new cross-ownership legislation, the privatization of state-owned terrestrial television networks, the deregulation of the cable and satellite television business, the privatization of telecommunication systems, and fundamentally the liberalization of foreign investment in media operations.[70]

This alliance between national and transnational capital provides a crucial basis for attempts to transcend prior limits on the culture industry's accumulation process. In complex and varied ways, in consequence, accelerated commodification reorients the institutional infrastructure through which the audiovisual trade flows. These efforts do not necessarily generate global uniformities in programming or in the specifics of national regulatory policies, but they work toward an overarching congruence, in that systems of provision are reshaped by an increasingly omnipresent capital logic. Let us turn to survey some of its characteristic modalities.

Throughout the large, developed market economies of Western Europe, Japan, and North America, this movement takes one major form. "One can recognise," Nicholas Garnham summed up the British case in 1983, "that what is at stake in the struggle surrounding the new information technologies is a battle between the public service and market modes of cultural production and consumption."[71] We already saw that, in the U.S. context, advocates of convergence and accelerated commodification took aim at public-service telecommunications; related attacks were waged on public-service broadcasting.

Public-service broadcasting was a development of the middle third of the twentieth century. Its mission and formal structure varied in relation to the strength of social class forces and the specific historical circumstances out of which it emerged, but in general, public broadcasting was constituted as a reformist alternative to what elites viewed as dangerous encroachments on the social order by radical movements of the working class. Taking shape in the interwar context, public broadcasting relied on different kinds of state-subsidy funding to establish a vital new institutional means of broadening access to cultural resources deemed essential for citizenship. As Graham Murdock explains, these included "comprehensive and accurate information about contemporary events and the actions of power holders; access to the contextual frameworks that convert raw information into usable knowledge by suggesting interpretations and explanations; and access to arenas of debate where contending accounts, aspirations, and positions can be subjected to sustained scrutiny."[72] Embodied variously in Canada, Germany, Japan, and a number of Asian and African countries, public-service broadcasting's foremost exemplar was housed in London. The British Broadcasting Corporation, like some of the other systems that drew on it as a model, represented a historically developing achievement that, during the postwar period, could be hailed by Nicholas Garnham as "a real step forward in the attempt to create a common culture."[73] Its originating limits and the episodic political and commercial pressures it faced, however, have rendered this achievement massively vulnerable—not least during what we may now identify as the historical runup to accelerated commodification, the 1960s and 1970s.[74] The BBC's rejection of an institutional and ideological realignment that might have placed it on the side of the democratic and antiwar initiatives then sweeping Britain was condemned by Garnham as a "betrayal" of public-service principles.[75]

Nations face differing historical specificities, and analysts will differ in evaluating the success of diverse struggles to extend common cultures through public-service broadcasting. Throughout much of Europe, however, between the 1970s and the 1990s, public-service systems underwent transformational change, as state television monopolies were curtailed, and competition with newly authorized corporate-commercial broadcasters flared.[76] Where corporate-commercial rivals already existed, in Britain, Japan, and other countries, their role was often enlarged. In this environment of intensifying commercial competition, the position of public broadcasting weakened. As pressures on public broadcasters grew intense and new claims were advanced as to how the common culture should represent its constituencies, the institutional framework within which public-service provision had been set has been comprehensively transformed.

In Italy, where the tradition of public-service broadcasting is structured not, as in Britain, in terms of a putative independence from the state but directly by national political parties, the rapid ascent to dominance of a corporate-commercial logic has nonetheless established a commonality. Private local broadcasting had been authorized in the 1970s, and it expanded rapidly at the national level. Media concentration became extreme, as three commercial television channels were allowed to coexist within a single conglomerate, Mediaset. The company's owner, Silvio Berlusconi, has twice become the political leader of his country and is thus well positioned to influence the fate of his rival, the state-owned broadcaster RAI. By 2004, under the stewardship of Prime Minister Berlusconi, Mediaset had built up a 44 percent share of Italy's overall TV market. On the other hand, RAI, still retaining 45 percent of the national audience, was to be partially privatized, on terms likely to disadvantage it further in its contest with Mediaset.[77]

Not always accompanied by the breathtaking chicanery of a Berlusconi, analogous processes have been widespread in other countries. National television systems that had been oriented toward some variant of the public-service principle, however limited, are being outflanked by corporate-commercial suppliers and/or infiltrated by a converging market logic. Massive new investments are needed to participate fully in digital-system development, while greatly increased funding reserves are also required to pay for programs whose prices are continually inflated as a result of competitive bidding between public-service and corporate-commercial broadcasters. Meanwhile, the public broadcasters' audience ratings are declining, while ideological attacks on them intensify; their ability to rely on license fees—annual payments by all TV users, which provide most of the wherewithal at the disposal of public broadcasters—is increasingly jeopardized. In many countries, budget cuts are an additional menace; in some, privatization looms.[78] Rather than renewing the mandate for national public-service provision, at this decisive moment European policy makers have trained their attention on bids to establish corporate champions to compete in the global market for culture.[79] In addition to the private broadcasters TF1, Mediaset, and Telecinco, regional audiovisual conglomerates such as Rupert Murdoch's Sky satellite consortia and Europe's Canal Plus were authorized to take control of multiple audiovisual platforms. In the United States, where public broadcasting was long delayed historically and, once commenced, was only accorded a frail and limited institutional basis, strenuous attacks by the Bush administration during 2005 further crippled the existing system. These attacks may have been animated by a conviction that public broadcasting is ideologically tainted and economically wrongheaded, owing to its partial reliance on government funds. More importantly, however, at this formative moment public broadcasting is being forestalled from playing a substantial part in the worldwide transition to digital television.

Throughout the countries of the global South, the historical circumstances confronted by the protagonists of the new capital logic for culture are quite different. Public-service broadcasting, where it existed, tended to be weaker and less adequate than its counterpart in the developed world. One of two other models tended to predominate: either broadcasting operated as a corporate-commercial franchise, as throughout much of Latin America, or it took the form of a more or less unaccountable, top-down state monopoly, as in much of Africa, Asia, and the Middle East. Accelerated commodification has taken hold here as well, despite this variation, further elaborating its expansionary capital logic.

Top-down processes of institutional change have been shepherded by foreign and domestic capital in alliance with state agencies; working people, not to mention the poor, have little say. Privatization, liberalized market entry, relaxed limits on consolidation and cross-media ownership—deregulation rather than democratization—are the results. Whereas, prior to the early 1980s, there had been virtually no non-state broadcasters in sub-Saharan Africa, by 1995 there existed 137 private operators in twenty-seven countries.[80] Government's role, however, did not consequently diminish. Pradip Thomas refers to a "control state" in the Indian context, where public-service broadcasting as a medium for nation building had earlier claimed a substantial presence—and the term "control state" might be applied much more widely, perhaps not only throughout Africa and Asia.[81] State power remains critical in granting increased freedom and material support to profit-making media services catering to advertisers via franchise awards, mandates of access to the electromagnetic spectrum, and loosened regulations. Like their counterparts in the United States and Europe, newly substantial conglomerates based in the global South, such as India's Tata Group, with its interests in computer hardware and software, telecommunications, advertising, and music, depend on the state to help propel convergence, further transnational initiatives, and manage the domestic culture market.

As corporate investment powers into television, channels have proliferated. As of early 2004, the fifteen countries of the European Union possessed thirty-eight conventional (analog terrestrial) public-service TV channels and forty-three privately owned conventional channels. The same group of nations now includes an additional seventy-five publicly owned channels with national coverage via cable, satellite, or digital terrestrial transmission, as compared to 702 privately owned national channels of this kind.[82] Movie and sports channels show the largest gains, but news, shopping, and childrens' channels have also increased significantly.[83]

This trend toward corporate-commercial channel development is not confined to Europe. In India, more than 180 channels were available by 2004.[84] Satellite dishes in the Middle East could access about fifty major Arabic-language channels and many additional European and American networks.[85] In the United States, by 2004 there existed 388 national nonbroadcast programming networks, of which cable operators had ownership interests in eighty-nine.[86] Gains by nonnational channels—that is, corporate-commercial services directed beyond domestic frontiers—are especially noteworthy expressions of this process.

Following along the path already cut by shortwave radio, regional and intercontinental television networking initiatives originated soon after the emergence of television.[87] Desultory movement toward "transborder TV" continued during the 1960s and 1970s, as dozens of countries joined the U.S.-dominated International Telecommunications Satellite Consortium (Intelsat); but satellite television transmissions that crossed national boundaries continued to "be channeled through local networks" and, with some exceptions, subject to approval by nation-state authorities.[88] However, a means by which to circumvent this obstacle to commercial expansion was moving from imagination to design and implementation: broadcasting by satellite directly to home receivers. Through the 1970s, the movement for a New International Information Order and continued U.S.-Soviet rivalry helped keep at bay such supranational satellite systems. By the mid-1980s, however, direct-to-home broadcast satellites were deployed throughout a few large-market nations and soon elsewhere; concurrent arrangements began to be made for satellite distribution of commercial networks—the first were forged, significantly, by the U.S. news channel CNN.[89] With these twin developments, the heavens began to smile on transborder services.

In 2004, no less than twelve thousand satellite video channels were beaming information and entertainment to the world.[90] With the collapse of the Soviet Union, telecommunications liberalization, and the rapid expansion of corporate satellite investment, supranational networks have begun to act as a pivot of system development. In small European nations like Ireland, Austria, and the Netherlands, foreign channels claimed in excess of 30 percent of the daily audience by 2002; even in the larger economies of Germany, France, and Italy, foreign channels constitute between 14 and 42 percent of the total offered by cable and satellite providers.[91]

"What are the political and theoretical implications of the apparent diversification of media ownership and the multidirectional nature of global media flows?" ask Yuezhi Zhao and Robert A. Hackett.[92] The question is vital, its answer complex. Most of the newly abundant commercial channels have no aspiration to cut free of prevailing property relations or the existing interstate system but aim instead to carve out a lucrative niche within their compass. In exceptional instances where transborder services do pose a perceived threat to the established order, efforts are made to claw them back into the dominative political economy.

Such a case occurred around the satellite news service Al-Jazeera, which began broadcasting in 1996 in the context of the authoritarian and corrupt states of the Middle East. Al-Jazeera was cited early in its career for offering a fresh and welcome alternative to the region's existing official broadcast services.[93] Al-Jazeera was sponsored by the Emir of Qatar, who continues to sustain the greater portion of its $40 million annual budget. Why? According to the journalist Hugh Miles, the decision to establish the service was a "wily" one: "The government is savvy enough to know hosting a massive American airbase in its heartland is never going to be wildly popular, but it is better that any criticism be aired in public."[94] During its first few years, Al-Jazeera even drew praise from the U.S. State Department for its critical stance toward Arab governments.

Al-Jazeera's success has stemmed from its unique access to the wars in Afghanistan, Iraq, its coverage of the Israeli-Palestinian conflict, its provocative talk shows, and its general editorial independence. But, with rapidly deteriorating U.S.-Arab relations, these same features have also opened it to pressure, first from Arab states and thereafter from the U.S. executive branch. Saudi Arabia mounted—and as of 2005 still maintained—a boycott on advertising with Al-Jazeera. Incensed that the network's "inflammatory" news was interfering with its campaign in Afghanistan, the Pentagon bombed Al-Jazeera's office in Kabul.[95] Still, Al-Jazeera's popularity only increased; by

April 2003, as it covered the war on Iraq, it had become the most-watched news channel in the Arab world, with forty to fifty million viewers scattered not only throughout the Middle East but in Indonesia, Europe, Canada, and the United States.[96]

The relationship of Al-Jazeera to processes of public-opinion formation within the Arab countries is beginning to command the attention of scholarly specialists.[97] For present purposes, however, the more vital issue is pinpointed by Hugh Miles: Al-Jazeera has "broken the hegemony of the Western networks." Not only has it garnered popularity with Arab viewers and grown to include over seventy correspondents in twenty-three bureaus on five continents; it has also become a paid supplier of otherwise-unavailable video footage from Afghanistan, Iraq, and Palestine to the major U.S. and British networks, from Fox News and CNN to the BBC.[98] In a region of surpassing strategic importance to the United States, the achievement of Al-Jazeera's unembedded reporting thus has been to break the embargo on independent news exercised internationally by western news agencies and networks.[99] The effect on the ability of the United States to shape and control information circulating within and outside this strategic region has been profound. "In Operation Desert Storm (1991), the American-led coalition was largely able to control the information war," writes Marc Lynch. In 2003, in contrast, the Americans "wanted to maintain information dominance but seemed powerless to achieve it."[100]

What is the larger meaning of this change? Is Miles correct to conclude that, "in the decades ahead we can expect only more Al-Jazeeras"?[101] Does Al-Jazeera presage a widening international movement by the media toward ideological diversity?

Parallels certainly could be found; let us briefly examine one seemingly congruent initiative. Televisora del Sur ("New Television of the South," or "Telesur" for short), was established in Venezuela in 2005 to provide a Latin American satellite news service. Telesur was accorded scantier support than that enjoyed by Al-Jazeera, a mere $2.5 million in seed money.[102] Support from the Argentinian government elevated Telesur into a joint venture, and additional aid—either money or technical support—was promised by Uruguay and Cuba. Telesur also signed a cooperation agreement with Al-Jazeera itself.[103] According to its director, after Telesur began broadcasting in July 2005 it would "wage battle in the mass field of television," to promote a "Latin American project of communicational integration."[104] The Venezuelan president Hugo Chavez himself declared that Telesur aims at "counteracting the media dictatorship of the big international news networks" like CNN.[105]

The political sources of Telesur must be set within a context fraught with social and political struggle. Bitterly opposed not only by the United States, which laid plans for its own counter-satellite service to broadcast into Venezuela,[106] but also by Venezuela's own capitalist class, Chavez has embarked on a prospectively far-ranging process of economic "indigenization," which redistributes some of the nation's substantial oil revenues to address the needs of the poor. As a result, international news media have painted him as a tub-thumping despot and Venezuela as a rogue state. Recalling Chile in 1973, where commercial media interests helped topple the democratically elected socialist president Salvador Allende in a bloody coup, Venezuela's corporately owned media (led by the Cisneros Group and Radio Caracas Television) acted complicitly with the coup d'etat that momentarily unseated Chavez in April 2002. After two days out of power, during which U.S. leaders quickly recognized Venezuela's new and illegitimate government, Chavez was borne back into office on a tide of popular mobilization.

The Chavez administration responded not only with Telesur but also by supporting a growing movement toward community broadcasting throughout Venezuela. Support for Telesur has come from governments that, for varied reasons, have distanced themselves from the United States. No matter how these struggles are resolved, they show that ideological diversity remains a prize to be won, not a natural condition or a default setting. In some places and at some times, transborder services may harbor emancipatory potential. However, within a global culture industry historically built and episodically reorganized in support of dominative social relations—a system thoroughly beholden to state and corporate power—diversity will not arrive as a passive by-product of audiovisual-market liberalization, or channel proliferation, or surplus telecommunications. Rather, it can come only by way of effective struggles for media access by sociopolitical movements actively challenging the dominant political-economic order.

This kind of mobilization remains exceptional. Further consideration of Al-Jazeera suggests additional grounds for skepticism that, if it hews to its present course, the culture industry will be redeveloped in accord with a truly diverse ideological template.

Pressures exerted against Al-Jazeera have not relented. Evidently enraged by its coverage, whose major sin has been to take scrupulous care to observe the "western" journalistic ethic of independence and impartiality, the Bush administration has rolled out rival propaganda services aimed at Muslim countries. The BBC also announced that in 2007 it would spend $34 million annually to begin broadcasting an Arabic-language television news and information service twelve hours a day across the Middle East.[107] Meanwhile, the United States also worked with unusual aggressiveness to manipulate the existing international and domestic news systems.[108] An "extensive, costly, and often hidden" information war has been undertaken by the Pentagon in alliance with other government agencies and private contractors such as the Lincoln Group.[109] Military force has likewise been repeatedly deployed. During the Iraq War, the United States bombed Al-Jazeera's Baghdad newsroom, killing a correspondent, and by April 2004, a total of twenty-one Al-Jazeera journalists had been held and released by troops in Iraq.[110] Al-Jazeera's reporters were eventually banned from Iraq.

A parallel initiative began to surface in 2005 and demonstrated that carrots as well as sticks were available to try to bring Al-Jazeera back into the fold.[111] With the appointment to Al-Jazeera of a British national who formerly worked for Associated Press Television as its new managing director, its launch of a sports channel and additional expansion planned around Arabic documentary and children's channels and Al-Jazeera International, an English-language news channel, and a commissioned study of how to privatize Al-Jazeera's expanding system by the consultants Ernst and Young, more supple means of co-opting ideological threats may be in the making. Miles speculates that, in the eyes of elites, Al-Jazeera provides a tempting commercial target:

> A more likely possibility than being blown up or shut down is that Al-Jazeera will one day be gobbled up by one of the world's giant media conglomerates. In the past network spokesmen have stated that any offer would be considered, as long as the buyer will guarantee editorial independence. . . . Besides, as long as the Saudis maintain the embargo on advertising, Al-Jazeera is not very commercially attractive. But the Middle East has a youthful population which holds enormous promise for advertisers. If something were to happen to the Saudi government or its advertising embargo, Al-Jazeera would suddenly shoot up in value.[112]

Whether privatization will occur, and whether advertisers and satellite and cable systems will sign up to support its English-language service, Al Jazeera International, remains to be seen.[113] At the time of writing, Al Jazeera continued to act as a principled defender of editorial independence. Still, it is judicious to remain mindful of historical precedents.[114] In keeping with these precedents, a move to corporate-commercial ownership could be clearly designed, in advance, not only as a profit strategy but also as a means of tempering Al-Jazeera's news coverage. "Plenty of potential investors in the region would like to see its coverage restrained," muses the *Financial Times*.[115]

Upon entry, or reentry, into the corporate-commercial culture industry, a thick web of existing structures and practices can be counted on to reshape and canalize overly intrepid or incautious programs and channels.[116] Supranational corporate-commercial TV infrastructures encompass audience ratings and related measurement companies, whose results are systematically employed to further capital's profit strategies over public service. Corporate advertisers pervasively influence programming choices to ensure that programs furnish a conducive ideological environment for their efforts at persuasion; there are numerous baneful indications that the supposed brick wall between editorial and sales departments, never as formidable as claimed, of late has been breached regularly. Political authorities also have long cultivated means of shaping the news to their taste. A rash of scandals in the United States confirms that, today, the press's complicity in its own manipulation has become profound.[117]

To equate the growth of transborder television with the cause of human freedom therefore is, in general, to miscast its importance. Early in 2000, Asia's satellite-delivered Star TV was watched in eighty-two million homes in fifty-three countries; its parent, Rupert Murdoch's News Corporation, enjoyed the capacity to reach 60 percent of the world's homes through satellite systems reaching five continents—Australia, Asia, Europe, and North and South America.[118] The communication scholar Daya Kishan Thussu shows how the program services that piggybacked on Murdoch's and other conglomerates' transborder satellites—MTV, ESPN, CNN, Discovery, BBC World, and a scattering of others—routinely amass the largest audiences.[119] As Miles points out, "[I]t is easy to forget that the most popular shows on Arab television are Egyptian and Lebanese soap operas and films, as well as imported Western game shows. *Who Wants to Be a Millionaire?* is one of the biggest hits of recent years, presented by a debonair Lebanese sex symbol."[120] Owing to the expansion of corporate-commercial media systems, the same point could be made with respect to almost any country.

The power of the culture industry does not translate directly into control of human subjectivity per se, and its assimilation into popular consciousness and experience remains a complex matter. And programming is also undergoing complicated processes of recomposition. Control over system development and programming and scheduling decisions, however, is gravitating not to viewers but to capital, directly via the new capital logic with which it is being suffused. For the moment, there turns out to be room for an Al-Jazeera. But one, two, or many Al Jazeeras? There is considerable reason for doubt, unless and until countervailing sociopolitical forces become strong enough to make the radical reconstruction of the culture industry a major objective.

Again, this is not to suggest that cultural programming is not undergoing complex changes. Corporate-commercial development and channel proliferation have fueled a surge in worldwide demand for audiovisual product, and this growth in turn has provoked apparently contradictory trends on the supply side. Local cultural production has gained visibility and at points real strength, even as U.S. cultural exports in particular and the role of transnationally organized distribution networks have expanded around it.

The U.S.-based culture industry has long been the paramount purveyor to the global market; during the 1920s and more extensively during the 1960s and 1970s, the extent of this skew and its strategic importance in reintegrating the global market under U.S. stewardship engendered widespread charges of threats to national sovereignty or cultural imperialism. This structural imbalance in the audiovisual sector has persisted. Exports by U.S.-based megamedia companies have surged, and Europe's negative trade balance with the United States in films, television programs, and video has increased; one U.S. report crowed in 2000 that "U.S. movies continue to dominate international trade in motion pictures, including film rentals . . . videocassette rentals and sales, and sales of television rights (including pay television)."[121] In 2000, the United States enjoyed an $8.1 billion surplus in its audiovisual trade with the European Union, divided equally between television and film rights.[122] Through early 2004, the ten top-grossing films of all time at the international (non-U.S.) box office were all U.S. films. In testimony to the new marketing muscle available to Hollywood, these were all recent blockbusters; the oldest of the ten—*Jurassic Park*—was released in 1993, and no fewer than six came out after 2000.[123]

However—and taking due note of substantial unevenness by country and by medium—"local" cultural production has also increased throughout much of the world. This still-fluid and complex movement to local production must be situated with respect to the culture industry's changing accumulation process. There are good reasons to suppose that it often denotes shifts in audiovisual product sourcing and therefore paradoxically constitutes a further development in the ongoing transnationalization of the political economy of cultural production.

Major advertisers and their agencies develop campaigns and purchase media time and space on a global basis; this involves placing pressures on media companies to act in concert with their supranational designs. Obligingly, the leading culture conglomerates themselves increasingly strategize and act on a transnational scale. But the advertising industry and the media system they depend on are reconfiguring themselves around segmented and differentiated audience groupings. Symptomatically, some—not all—efforts at "local" cultural expression come in the form of such audience-segmentation initiatives. In 1999, Sony expected to create no less than four thousand hours of TV programming in foreign languages—more than twice its output of English-language programming—for its own foreign channels and for sale to other groups.[124] Sony produced thirty-three hours a week of Hindi-language programming and transmitted it not only to India but also to Africa, Britain, North America, and the Middle East; it also produced or coproduced thirty-five weekly hours of Mandarin-language programming, mostly for its Super TV channel that reaches most Taiwanese households, but with an eye on mainland China.[125] Time Warner is increasing the number of non-English-language films it produces from thirty-two in 2005 to forty-four in 2006.[126]

What passes for local culture is, in addition, selective and perhaps newly malleable. Within the United States, where laws capping group ownership of radio stations at forty nationwide were eliminated in 1996, Clear Channel Communications was operating over twelve hundred stations by 2002. A dominating presence among the handful of giant new radio chains, Clear Channel has moved quickly to "perfect the art of seeming local," deploying a technology called "voice tracking" to customize programming. By piping in voices and inserting references to local events, people, and landscapes, the technology makes it sound to listeners as if each station's disc jockeys actually work within their community; in fact, the same songs are being played on stations with similar formats nationwide.[127]

Centrally orchestrated "localization" strategies are now a staple in a growing range of geographic and media-product markets. The largest culture-industry conglomerates, wrote a business analyst, "may well maintain their leadership roles, but they will have to adapt to local audiences and include local content. Moreover, foreign content will begin to nudge its way onto the global hit parade."[128] "What we're seeing," concluded an observer of the international film industry, "is the emergence of a two-tiered global system. English is the language of the international blockbuster, but lower-budget pictures can be made in almost any language for the home market, and a few . . . will even become international hits. Hollywood, with its vast corporate resources, can call the shots in both tiers."[129] "'Twenty years from now,' says Time Warner's CEO Richard Parsons, this two-tiered model "'is going to proliferate across the spectrum of things we do.' Doing it right, he adds, 'is probably the biggest challenge facing the company.'"[130]

A growing fraction of local programming testifies to innovations in corporate structure and function, which link transnational and national capital in a warren of coproductions, joint-ventures, and other partnerships. Ownership changes and new financing patterns have likewise gripped the command centers of the global system of marketed cultural provision.

"Hollywood" itself, long a symbol of U.S. global cultural domination, has been substantially transnationalized in structure, even as it continues to parlay its dominance of the U.S. audiovisual economy and its control over global distribution into a commanding share of the world market. Only one of the six major U.S. film producers is owned by a non-U.S. culture-industry conglomerate: Columbia (Sony). But two others have shifted out of U.S. ownership at least once over the preceding fifteen years: Fox, owned by News Corporation, altered its legal home to concord with the preferences of its major owner, the Australian Rupert Murdoch; and Universal moved from U.S. to Japanese to Canadian to French and back to U.S. control. Foreign capital has likewise flooded in to the second-tier units of "domestic" U.S. cultural production. A series of allied structural changes have contributed to the same trend.

U.S. film studios routinely presell the foreign rights to a film before production is authorized. "For a $50 million film, a U.S. producer might seek 10% to 12% of the total budget from a distributor in Germany, 7% to 9% from France, 6% to 8% from Italy, and 4% from Spain, amounting to nearly a third of the movie's budget from the major continental European countries."[131] Across the matrix of preproduction, production, and postproduction activities, shifts are also occurring in the location and function of audiovisual production. "Global Hollywood" constitutes not merely a U.S. phenomenon but the activating core of a "new international division of cultural labor"—an unequally shared and dominative projection by cultural capital, especially transnational capital—at large.[132]

While the huge U.S. domestic market—far bigger than any other national market, and even bigger than the combined markets of Europe, the Middle East, and Africa[133]—continues to exert sustained gravitational pull over the roiling global audiovisual system,[134] its role is also changing. Partly akin to what are aptly termed "Japanese-owned but U.S.-oriented multinationals" like Nintendo and Sega, the French video-game makers Ubisoft Entertainment and Infogrames Entertainment "publish Americanized games, rely on

North American game designers, and are very dependent on the U.S. market for their sales."[135] As in other consumer-product sectors, the United States culture industry is functioning increasingly not as a forum for the expression of national culture but as a pivot of global demand.

In the emerging political-economic system, the role of the conglomerates is thus increasingly to pool transnational capital, to produce commodities within the new international division of cultural labor and propel them outward into the world market—at least when they are deemed to possess a lucrative crossover potential:

> While the U.S. still produces many of the latest models, more and more it serves as a kind of cultural chop shop, retooling offerings from afar. It has a lot to choose from, as there has been an explosion in both the quality and volume of entertainment generated abroad. The privatization of the television business in Asia, Europe, and Latin America and the advent of cheap production technology has helped spawn a generation of ambitious young artists and producers who believe they can conquer the world. . . .
>
> The big U.S. entertainment conglomerates, and the foreign-owned companies like Sony that run their entertainment operations almost entirely from the U.S., are embracing this phenomenon. The companies found in recent years that it was getting tougher to just jam American-made product down the world's throat—as when MTV learned that it needed to mix in more local acts on its international channels. So in response, the media giants are increasingly flipping the equation on its head, scouring their overseas operations for talent that can be buffed up for the big U.S. market. Success in the U.S., in turn, can open the door to even greater success in the rest of the world.[136]

Repelled by slowing growth rates in the United States, tempted by higher ones in China, India, and select other countries, and always trying to spread their risks across multiple media and product markets, major U.S.-based culture conglomerates continue to shift away from simple exports of U.S. cultural commodities. Exports, as the *Financial Times* delicately underlined, "particularly if the current geopolitical climate persists, will not be a long-term grown engine. Instead, companies have to take their business models and, in effect, dress them up in local costume."[137]

There is no question, therefore, that the culture industry is altering in light of proliferating product markets and transnationalization.[138] It is becoming more multipolar: units of capital domiciled not only in the United States but also in Japan, Western Europe, South Korea, China, Mexico, India, and other nations have acquired substantial roles, sometimes even leading roles, in the continuing process of commodification. Although I have argued that this changed structural foundation does not itself propel a new ideological openness, at least with regard to the question of capitalist control of the political economy, it remains important to ask whether it affords a greater measure of cultural pluralism. Does increased multipolarity vitiate the long-standing U.S. dominance over the system of global cultural provision? Does it signify that the United States is losing the ability or inclination to put culture to imperial ends?

On the evidence of the period since September 2001, the inclination has not diminished; however, U.S. dominance faces some new political challenges. General endorsement of a UNESCO declaration in favor of protecting cultural diversity constitutes one recent counter-thrust by the international community; suddenly pointed opposition to U.S. power over the global Internet represents another. These challenges need to be situated briefly in their historical and conceptual context.

Control over international cultural production and distribution constituted a prize wrested from Great Britain by the United States during the first half of the twentieth century. In multiple ways, analysts have long known that this control contributed in crucial ways to the rise of dominance of the United States. As I and my co-authors put this in 1992, "From the ability to frame the international news agenda, to the deployment of high-technology weaponry, to the cultivation of an unremittingly commercial ethic of mass consumption, communications has become an essential dimension of power."[139] These functions continue to be performed. But the U.S. government's strivings to function as the chief steward of the culture industry's transnationalization occur within a profoundly changing context.

Power to develop culture as a commodity on a transnational scale has transferred over the course of a generation to corporate capital, as privatization and commercialization have proceeded and as new distribution systems have been consecrated to the circulation of commodities. At these points and others, commodification has been expedited and sustained not merely through the exercise of economic power but also—sometimes preeminently—through political power. That states defer so often to capital, and even that national sovereignty is continually transgressed, thus does not signify that states have ceased to be important. Far from it. States remain crucial to the project of commodification, within the unfolding context, as Ellen Wood emphasizes, of capitalist transnationalization.[140] "The very fact that 'globalization' has extended capital's purely economic powers far beyond the range of any single nation state means that global capital requires many nation states to perform the administrative and coercive functions that sustain the system of property and provide the kind of day-to-day regularity, predictability, and legal order that capitalism needs more than any other social form," writes Wood; today, therefore, the global political economy is "administered by a global system of multiple states and local sovereignties, structured in a complex relation of domination and subordination."[141]

This process is anything but conflict-free. States confront not only different interests and environing constraints but also distinct "internal needs and pressures" as they negotiate with other states.[142] Growing economic multipolarity generates significant added potential for political friction. During recent years, international political conflicts over culture and information have been sharpening.

In deliberations over trade in cultural commodities, the transnational culture conglomerates continue to demand, as they have for decades, that for policy-making purposes audiovisual trade should be treated like other cross-border commercial transactions.[143] Any kind of barrier or constraint imposed over national audiovisual space, they insist, is obsolete, unjust, and even irrational. Their initiatives to promote the capital logic of transnational commodification, either via monocultural blockbusters or by selectively appropriating and exploiting cultural difference to segment audiences, must be untrammeled. The free flow of cultural commodities—*their* commodities—should not be interdicted.[144] Again confirming a long-standing pattern, these companies have been strongly supported by the executive branch of the U.S. government. One strategic objective is to limit and, if possible, to reverse cultural exception clauses mandated by existing trade treaties (of the General Agreement on Tariffs and Trade and its successor, the World Trade Organization, and, regionally, of the North American Free Trade Agreement)—clauses promulgated in the early 1990s, putatively to protect national cultures, in response to a prior phase of U.S.-led commodification.[145]

On the other side of this issue stands the rest of the world: the European Union, Canada, numerous African and Latin American nations including Mexico and Brazil, and, interestingly, Japan and India as well as China. In 2005, these states officially endorsed a plan to strengthen the protection of cultural diversity through a legally binding UNESCO convention, which they hope will be accorded priority over the jurisdiction claimed by the World Trade Organization. The United States position opposing the effort to defend cultural diversity was backed by only one other nation—Israel—while no fewer than 148 countries approved the "Convention on the Protection of the Diversity of Cultural Contents and Artistic Expressions."[146]

Although the convention on cultural diversity does not comprise an incendiary document, although it does not supplant the dispute-settlement mechanism offered by the World Trade Organization, and although it will likely exercise at most a modest restraining influence rather than a more basic form of control over commodification, the battle was joined—and the United States lost by an overwhelming margin. This result occurred despite the fact that the United States seems to have expected political support to issue from nations that now host significant units of culture-industry capital: Mexico, Brazil, South Korea, Japan, and India, for example.[147] Nations backing "cultural diversity," however, do so out of disparate motives: some based on a principled conviction, some attempting to secure commercial advantages for domestic capital. In this instance, growing economic multipolarity translated into a significant political reversal for the United States within the interstate system.

Might this reversal portend a tectonic change? A suggestive indicator derives from initiatives pertaining to Internet governance, initiatives that culminated at a 2005 World Summit for the Information Society (WSIS) conference in Tunisia.

International anxieties about U.S. power to shape and control the global Internet have formed a subterranean policy current for years.[148] One partial expression of this concern are some of the various initiatives to establish alternative domain-name authorities outside the purview of the U.S. Internet Corporation for Assigned Names and Numbers (ICANN), which runs the master address system used to direct traffic on the global Internet.[149] During the summer of 2005 these fears about U.S. power over the online world crystallized. Late in June, notwithstanding appeals from various countries for a new multilateral governance structure for the Internet, and reneging on a Clinton administration promise that it would bow out of the picture, the U.S. Department of Commerce announced that it would indefinitely retain its existing oversight authority over ICANN.[150] Then, in August, apparently for the first time, the U.S. government interceded overtly with ICANN: the Commerce Department asked that ICANN delay its approval for a much-bruited ".xxx" domain name.[151] This act offered fresh evidence, even to nations in agreement with the Bush administration that "adult" Internet sites should not be accorded any added legitimacy, that U.S. veto power over ICANN constitutes an unacceptable transgression. As the *Chicago Tribune* explained in an editorial, it also "put the U.S. in the awkward position of opposing more international control of the Internet, while engaging in its own political meddling on the Internet."[152]

A concurrent initiative by the International Telecommunications Union, an agency affiliated with the United Nations, offered a proximate opportunity for redress by the international community. Most of the action actually unfolded during the weeks leading up to the World Summit on the Information Society convening in Tunis in mid-November 2005. China, Iran, and other developing nations renewed their arguments in favor of opening up control over the Internet to wider participation by non-U.S. states. But a crucial realignment came in October, when the European Union unexpectedly proposed a "cooperation model" that would grant countries besides the United States a formal role in guiding the Internet domain-name system and governing Internet policy.

Howls of protest at this betrayal issued from the U.S. press. These are noteworthy, first, for their undisguised assertion of a U.S. right to preside over the Internet's global jurisdiction. A *New York Times* op-ed declared that, on economic and national security grounds, "the United States should keep control of the Internet": "The Web of the Free," it trumpeted, is rooted in "American values."[153] "Mitts off the Internet," added a writer in the *Chicago Tribune:* "Is the Internet so broken that it needs to be fixed by the likes of Iran, Cuba, China, Ghana, and France?"[154] Taking aim specifically at "our friends at the European Union, who last month turned against the U.S.," the *Wall Street Journal* nonetheless portended that "Washington doesn't hold all the cards here."[155] The Information Technology Association of America likewise singled out the European Union for "compromising Internet governance in catering to China and other smaller countries" through its "anti-business position."[156] Senior U.S. lawmakers snapped into action as a bipartisan group of congressional representatives wrote to the U.S. Departments of State and Commerce to urge that, by refusing to cave to international pressure, "the United States should maintain its historic role."[157] The issue of Internet governance then ascended to the topmost ranks of the country's political leadership. "In a sign that traditionally obscure discussions about Internet control have taken on new prominence," a story confined to a specialized online news service reported, "President Bush broached the topic in a White House meeting . . . with European Commission President Jose Barroso."[158]

When the dust settled after the WSIS deliberations in Tunisia, the hardline U.S. policy had prevailed; the United States retains its grip on the day-to-day management of Internet domain names via ICANN. But the status quo in fact has been altered. In deference to the international community, though barely noticed by the U.S. press, a multifunction "Internet Governance Forum" was also established. U.S. leaders crowed that they had won the day, but a European Union spokesman could declare, equally validly, that the agreement "represents a clear commitment to a more multilateral and transparent approach to internet governance."[159] Issues associated with the control structure and policy of the global Internet remain in flux and very much alive. The possibility is real that growing economic multipolarity will translate into significant political change.

Even as it is reorganized around more multipolar political and economic axes, how is the capital-logic of the transnationalizing culture industry concurrently altering in the vortex of technical change?

Toward a Universal Product Market for Culture

By 2006, news stories heralded, convergence had finally become palpable: television is being conjoined with computers, wireline, and wireless broadband networks to engender a raft of new devices and program forms—video on mobile phones, iPods, TiVos, Web sites, blogs. This "bubbling digital stew" in turn works to "give consumers more control." Ted Leonsis, vice chairman of America Online, elaborated: "There is this primordial digital soup brewing of more bandwidth, more storage, more devices, and more people creating content. . . . The lightning that struck is that the people have rapidly adopted all this even faster than we in the industry conceived, and bypassed the traditional media."[160] In this new "anything, anytime, anywhere paradigm," as the media impresario John Malone called it, everything is up for grabs.[161]

Everything? By opening the doors to profit-seeking investment in networks—including terrestrial, cable, and satellite broadcasting—capital obtains access to tempting new tracts. Internet systems and services and mobile telecommunications, each progressively augmented by broadband capabilities, comprise further investment nodes. Although this process of reorganization is complex and wide-ranging in scope, and though it remains volatile and incomplete, it continues to be shaped by the abiding principles of the capitalist political economy.

The genres and practices that will animate a digital culture industry are in flux, and the technical basis for this massive reorganization of cultural production and distribution has not yet stabilized.[162] As a condition of their continuing self-assembly in this dynamic context, the conglomerates are attempting to absorb and assimilate digital technology and program forms. Thus commences what is certain to be a long-term metamorphosis, replete with contentious and complex political-economic and cultural elements. Again, however, as against the gamut of market-churning consumer-electronics hardware and network infrastructures and the eruption of novel expressive genres and forms of information access and exchange, we must not lose sight of the emphasis on which corporate conglomerates are themselves intransigent. From the standpoint of capital, new technologies and services offer mere means.

The end they serve is the enlargement of profitable revenue extraction. This takes two major forms, with two others that might become more consequential: advertising, or selling commercial access to audiences and users; selling content and services to institutions and individuals via a practically endless variety of fees, rentals, site licenses, and subscriptions; data sales; and transaction charges (think of Ebay). As the Internet has emerged as the supreme incarnation of convergence on a global scale, it has likewise become the overarching object of commercial desire. Over the course of a single decade, the Internet has been transformed from an utterly noncommercial to a pervasively profit-oriented system.

Again, however, the present is more deeply rooted than is often appreciated. By the 1970s, it was already evident to U.S. policy makers that, owing to their great economies of scale—"high fixed costs and low variable costs"—database services were likely "to exhibit tendencies toward monopoly." Ithiel de Sola Pool and Arthur B. Corte projected this trend: "It seems clear that worldwide need for information will be met by remote access to computer databases; that in many instances there will be only one or few large databases in a field; and that, therefore, the countries and organizations that take the lead in creating the databases and the communication system for accessing them will gain substantial influence in world affairs."[163] Is it mere coincidence that, by around 2000, fewer than 5 percent of the world's publicly accessible databases were produced in Asia, Africa, the Middle East, and South America?[164] Is the prominence of Google and other U.S.-based Web companies—whose existence could not have been imagined in 1975, but which in 2004 made up all of the top five Web destinations in Europe—simply fortuitous?[165]

Much cultural and political-economic work was needed to effectuate this result. The reconsecration of the Internet as a commercial system has occasioned repeated intervention as well as important attempts at coordination by elites. In Paris in September 1999, twenty-nine of the world's most powerful technology and media executives, including Steve Case of AOL, Thomas Middelhoff of Bertelsmann, Louis Gerstner of IBM, Michael Eisner of Disney, and Gerald Levin of Time Warner, met behind closed doors "to debate how to turn the wild Web into a family-friendly place to do business."[166] Many of these executives (all five of those named above) have retired or became casualties of the media and technology bubble; their visions of how to commercialize the Internet competed and clashed as the structural interests they represented fought for market dominance. Yet the goal they shared endures: A dynamic Internet infrastructure has been increasingly colonized by and for capital.

At a basic social level this endeavor has been thoroughly, even necessarily, uncreative. Leading features of today's Internet have been recast to act merely as extensions of long-standing corporate-commercial practice. First, the Internet embodies and further extends the tendency toward transnationalized corporate production and distribution. Second, the Internet simultaneously supports a deepened effort to market to consumers by differentiating and segmenting them into target groups. Third, as a "control utility," as Darin Barney labels it, the Internet sustains a trend toward more comprehensive corporate monitoring and metering of transactions.[167]

New digital cultural commodities (videocassettes and DVDs) and electronic distribution channels (satellites) have been deployed strategically for a generation to evade or end-run around existing regulations and restrictions by forging new modes of access to national markets worldwide. In 1996, as the Internet was exploding as a world-scale commercial system, a top organizer and strategist for the U.S. transnational audiovisual industry, Jack Valenti, called the emerging media "the new centurions of the digital age, marching over continents and across geographic borders, breaking down artificial government barriers, the most powerful audiovisual armies ever known."[168] Akin to other new media, the Internet has provided capital a long-desired means with which to press forward with its accumulation strategies, offensive and defensive, across a much enlarged technical, geographic, cultural, and political range. Its signal achievement has been transcending prior limits to the transnationalization of cultural production, distribution, and consumption, enabling nationally denominated markets to be further reconstituted and national political accountability to be further undercut.

Diversified cultural conglomerates and their advertiser patrons and technology suppliers also have been empowered to make the World Wide Web a more effective sales apparatus—by turning it into a self-service vending machine of cultural commodities. The prize is not a liberated zone beyond the market but merely new and effective channels through which to target and track desirable groups of shoppers. Even as traditional broad-reach media, especially print media, flail and flounder in search of commercial support and give up any pretense of public service, the Web is claiming increasingly substantial corporate advertising expenditures—around $10 billion in the United States by 2004.[169]

Well-capitalized companies rather than upstarts continue to have the edge in bankrolling carefully calculated versions of diversity. Internet navigation services, reports the U.S. National Research Council, "open an international audience to purveyors of content and services, no matter where they may be located."[170] They do so, however, not only via a process of "monetized search," in which search results may be sharply distinct from those produced by the carefully neutral practices of conventional librarianship, but also by extending the process of commercial localization from above. Yahoo possessed operations in twenty-two countries and thirteen languages by the spring of 2000, with 120 million users (and 3,500 advertisers) worldwide. The directory/portal boasted no less than 350 content partnerships across Europe alone, helping it to offer country-specific content and advertising.[171] By November 2004, Yahoo operated portals customized for thirty-two national and/or language groups.[172] Google's search engine was by this time even more accessible linguistically, permitting navigation to proceed in over a hundred languages and dialects.[173] How soon will one or the other of these engines of commodification pose a political challenge, under the rules of the World Trade Organization, to libraries as a public-service institution?[174]

In the heady days prior to the popping of the Internet financial bubble, enthusiastic projections of the Internet's transnational commercial potential were frequent; according to one writer, two-thirds of all e-commerce spending by consumers soon would take place outside the United States.[175] Another commentator suggested that, by 2003, more than half of the content on the Web would be offered in a language other than English, up from 20 percent in 1999.[176] These estimates proved to be inflated. The dominance of English as a global language is due to more than only the Internet. English-language Web pages still accounted for over two-thirds of the total counted by one survey in 2001, and although the preponderance of English was declining, as David Block underlines, "[G]reater diversity does not necessarily mean that all languages are equal: bigger is still better in the pecking order of world languages as much of the proportional weight wrested away from English has been in favour of a few major national languages." The top eleven languages represented by Web sites counted in 2001 accounted for 96 percent of the total, while only 9.3 percent of the online population of 619 million counted in 2002 did not fall within the compass of these eleven language groups.[177]

Content providers have for several years concertedly sought to develop "localization" market strategies. Discovery Networks' Argentinian Web site, boasted an executive, posted "'an article on the condor which was sourced in Spanish in Argentina but has also been translated into Portuguese for the Brazilian site, and we're hoping to offer that back to the U.S. site. . . . [Y]ou will not see us act as some companies do in the sense of rolling something out from a U.S. headquarters for the world.'"[178] Mainly to reach the huge Chinese diaspora, in 2005 Apple and Universal Music announced that more than a thousand songs by top Chinese musicians would be made available for downloading from Apple's iTunes stores in fifteen North American and European countries.[179] Darin Barney sums it up this way: "Global cultural and media giants have quickly gobbled up internet content and carriage enterprises because they understand that whatever threat the internet's technicality might pose to the reach of mass culture and the integrity of the mass audience, this threat signals a much greater opportunity to increase and diversify their hold on the cultural universe."[180] Yuezhi Zhao and I have put the point more sharply: capital is learning to parasitize cultural difference in order to profit from it.[181] A considerable further advantage of this market-development strategy is that it may help to deflect charges of cultural imperialism.

The Internet has carried to a new level a general tendency toward invasive corporate monitoring and metering of audiences/users. Digital media of many kinds are converging on social interaction across a fantastic range; as they do so, data are produced about that interaction. The data are then, increasingly, fed back into the system to engender what Kevin Robins and Frank Webster, nearly twenty years ago, presciently envisaged as "cybernetic capitalism."[182]

Conclusion

Transnational market development has made at least some room for the expanded circulation of local and vernacular cultural forms. Not all of this effort will or can be canalized by capital's designs. Intercorporate rivalries in securing popular programming and increasingly extensive market experimentation engender openings for at least some alternative programs. The technical skills needed to produce for the new media, though segmented by dominative class, gender, and race relations,[183] are not monopolized by the culture industry; Internet-based means of production are not the preserve only of wage labor. The declining cost of digital production continues to open avenues of expression to individuals and groups effectively barred from the commercial mass media. The replacement of U.S. stewardship by some agency of internationalized control over the Internet remains a simmering political issue. Although efforts to contain and suppress these countertendencies should not be underestimated, corporate-state control over the Internet remains thankfully incomplete.

In our emerging era of digital capitalism, struggles for justice persist. But they do so in the context of a still expanding and globally powerful culture industry. In the wealthy countries, and to a lesser extent elsewhere, backers of commercial interactive services are rapidly grafting them onto cable television, landline telecommunications, and wireless media systems. Dialup modem-based access limits the speed and constancy of service; thus it constrains the market-development process. To get around this—at least for most-desired groups of well-heeled consumers—culture-industry companies are building out capacious, high-speed, broadband PC access within the home, on tap day and night. The Internet is at the threshold of a further institutional metamorphosis, whereby it will transition into a control mechanism for digitally reconfigured devices and services of every kind, from washing machines and ovens and toasters to finance, power, lighting, and water. Even advocates of convergence have not yet widely recognized the sweep of this process. Corporations based in many different sectors—not merely in the audiovisual area—have deployed information and communications technologies to enlarge their scale and scope.[184] Adjacent initiatives, chronicled in an endless bombardment of business news reports, bid fair to transform the market into a 24/7 operation, anywhere and everywhere. Mobile Web-access devices allow the continued usurpation of sites outside the home, prominently including automobiles and airplanes, public spaces and shopping malls. The culture industry is thereby enabled to build—though only by continually tearing down or "repurposing" what already exists—an almost innumerable array of commercial platforms. The unfolding of the culture industry throughout daily life is the subject of chapter 6.

7

Parasites of
the Quotidian

"[I]t is never safe to conclude that puffing has reached its maximum distention," asserted Raymond Williams more than forty years ago.[1] Just so: more than a century after the advertising industry took over the job of fronting for a new brand of consumer capitalism, today the sponsor system is experiencing a renewed developmental surge.

How can this be? What fields are left to conquer? Advertiser sponsorship constitutes one of the basic business models on which today's privatized, deregulated culture industry is predicated; the main other model is direct sales and rentals of programs and services to consumers. Both models have been increasingly widely employed to preempt earlier forms of cultural provision based on government funding. Advertising possesses considerable room to expand, owing not only to newfound technological and institutional capabilities but also to the continuing rollback of prior limits.

Whole fields of practice funded and operated largely by state and non-profit agencies long restricted advertising, for example, in education and much of public broadcasting. Other segments of the cultural field, notably art museums and public libraries, were rendered inaccessible to corporate-commercial inroads because of their long-standing role in socializing elites within processes of class reproduction. Within divided societies, vernacular and informal cultural practices associated with workers, the poor, and racial and ethnic minorities often flouted middle-class sensibilities; by turns feared and disdained, these sometimes became objects of suppression. Only haltingly and selectively were they appropriated and transmuted into public imagery by the sponsor system.

Was this rollback a function of the advertising industry's sheer institutional power? Escalating advertising expenditures seem to suggest as much. Global advertising spending increased sevenfold between 1950 and 1996, growing one-third faster than the overall world economy. In absolute terms, staggering sums were allocated. At the height of the Internet bubble in 1999, world advertising revenues were estimated at $429 billion.[2]

However, absorbing the punishing effects of the Asian financial crisis and the deep downturn of 2001, when U.S. ad spending declined for the first time in forty years and by the largest percentage since the Depression year of 1938, industry growth slowed.[3] It did receive further boosts from corporate marketing tied to the Olympic Games and from television advertising outlays for elections in the United States and Japan. Global advertising expenditures for 2004 were expected to break the half-trillion-dollar mark for the first time (reaching $519.4 billion), with the United States accounting for just over half of the world total ($263.3 billion).[4] Between 1992 and 2004, moreover, after accounting for inflation, the U.S. population doubled its household debt to more than $10 trillion.[5] But this ballooning indebtedness testifies not only to the success of the sales effort but also to encroachments on the standard of living that made it harder for many to avoid running up their credit-card balances.

Underlying these complex quantitative trends are propulsive shifts in the institutional character of the sponsor system. These changes have come not only as a result of that system's formidable institutional strength but also, paradoxically, because of the fresh uncertainties and difficulties with which it has been forced to contend.

Advertising's Role in Consumer Capitalism

Selling constitutes a basic imperative of the market system—a necessary step in the cycle of capitalist production. The corporations that underwrite the system that has evolved to perform the selling function gain, in return, a commanding role in shaping and coloring not only advertisements but the larger culture. For these companies, it is a splendid bargain—especially when, as in the United States, advertising is treated for tax purposes as a deductible expense.

Capitalism's endless conveyor belt of new commodities begins with corporate research and development, continues through production and distribution, and culminates in an ever intensifying sales effort; these functions intermingle and even merge as capital's compulsion to subordinate consumption to production intensifies. Kraft Foods budgeted $800 million to fund the marketing of a hundred new products between 1999 and 2001, products that it hoped would account for most of the growth in its overall sales volume.[6] Gillette's introduction of a three-bladed razor for women (Venus) rested on around $300 million in research and development and manufacturing outlays; the company announced that it would spend $100 million in advertising, plus another $50 million in "first-year marketing expenses," to produce and distribute ads in twenty-nine countries.[7] Novartis, whose pharmaceutical business's top executive came there from Pepsi-Cola, planned to spend $1.2 billion over two years launching only five new drugs, while Merck's Vioxx relied on nearly $50 million in direct-to-consumer advertising between January and July 2004—two months before the $2.5–billion-per-year-generating painkiller was pulled from the market in a scandal over improperly ignored side effects.[8] Procter and Gamble's Crest products budget around a quarter of a billion dollars for media campaigns.[9]

The sales effort is coordinated by a shrinking group of advertising agency "super-groups." Through a protracted series of buyouts and mergers, four super-groups housing dozens of once-independent agencies were assembled: WPP, based in London; Omnicom and Interpublic, based in New York; and France's Publicis Groupe. Dentsu, headquartered in Tokyo, is also a major player, but it derives only 5 percent of its revenues from outside Japan; each of the super-groups, by comparison, garners between one- and two-thirds of its revenues outside its home market. The trade journal *Advertising Age* suggested that this "transformation of agencies from fiercely independent entrepreneurial enterprises to publicly held mega-giants comes in response . . . to marketer consolidation. Most clients today demand that their agencies offer global reach and seamlessly integrated communications capabilities. Those shops that don't face a choice: shrink or sell."[10]

As media corporations diversify to keep up with the proliferation of new media and the increasing fragmentation of audiences into differentiated niches and segments, they encompass a widening array of selling venues, from magazines to television to the Internet.[11] This process of industrial reorganization is a cause of and response to the fact that it is becoming increasingly difficult to engage consumers' attention.

Using "convergence" as a buzzword, the media behemoths emphasize to advertisers what they call "integrated marketing," whereby they offer to help sponsors target audiences with a cohesive selling message across once-discrete media units. To rebalance the terms of trade, in response to this consolidation among media companies the ad-agency super-groups have begun to take over what had been an independent "media-buying" function. Super-groups have acquired many of the specialized businesses whose function is the bulk purchase and subsequent brokerage to advertisers of time (for the electronic media) and space (for print media). The ad-agency groups hope that by integrating the media-buying function they can gain renewed power over the placement and cost of commercials and more generally over the programming in which they are inserted.

Each super-group derives around half of its multibillion-dollar annual revenue from services other than the design and placement of commercials: public relations, direct marketing, sales promotion, multicultural marketing, health-care marketing, interactive marketing services, consulting and corporate branding, and market research.[12]

Advertising agencies continue to make much of their ability to offer one-stop advertising and marketing services to major clients. Interpublic, hired by Coca-Cola to help define the "brand essence" of its leading soft drink, intended to "act as creative consultant and idea generator at the global level, evolving core messages about the Coca-Cola brand that could be incorporated in marketing locally."[13] The spirit of this enterprise accords well with strategic thinking that emphasizes the establishment of ongoing "brand relationships" with targeted customers. "It is our job to understand the relationship of products to people's lives, how they think and feel about our clients' products and how best to communicate the relevance of those products to create competitive value," declared Allen Rosenshine, a top agency executive. To accomplish this, "[W]e have to be media neutral in our planning and capable of coordinating and executing creatively in any channel of communication. . . . [W]e must provide convergence at the strategic planning level of product and design, mass and interactive advertising, direct marketing, sales promotion, packaging, point-of-purchase display, PR, infomediaries in one-to-one marketing, and all other yet-to-be developed forms of brand intelligence and communications."[14]

None of this, however, ensures that the triadic sponsor system will smoothly manage the transition to what has turned out to be a jagged and volatile selling environment. In a context of secular overproduction and transnational competition, individual consumer product manufacturers face added difficulties in profitably selling commodities; even the vaunted brand builder Coca-Cola has struggled to regain its corporate footing.[15] Advertising agencies, even as they bulk up into super-groups, have not managed to avoid turbulence. Agencies sometimes voice desperation as they try simultaneously to diversify beyond broadcast television, magazines, and newspapers into new media; to cut costs; and to keep up with rapid zig-zags in media use and cultural practice.[16] Media companies are also undergoing complex and often precipitous transitions. In the United States and to an extent elsewhere, broadcast television, newspapers, radio, and magazines are being hit with varying force by their advertiser-patrons' growing preference for more targeted, and often newer, media: the Internet, cable television, and direct mail.[17] During the 2005 "upfront" negotiations between advertisers and broadcasters, when sponsors make advance purchases of as much as 85 percent of the TV networks' inventory of commercial time for the next year, drug companies like Pfizer and consumer-products behemoths like Procter and Gamble—the leading U.S. advertiser, spending around $2.5 billion on TV—cut back on TV commercial buys in favor of other media.[18]

Institutional transformation and intensifying competition do not end here. The buildup of proprietary communications systems by big companies of every kind already comprised a noteworthy trend twenty-five years ago, and it has markedly deepened.[19] Wal-Mart, whose legendary computer-communications systems gave it unparalleled ability to refine its selling mechanism by mining and sifting and exchanging proprietary data, in 1998 added to its arsenal a Web-based TV network throughout most of its 2,600 stores. By 2005 Wal-Mart TV reached 130 million viewers every four weeks, making it the fifth-largest TV network in the United States after NBC, CBS, ABC, and Fox. The network in effect competes for advertising with these and other media companies by showcasing for shoppers "a constant stream of consumer product ads purchased by companies like Kraft, Unilever, Hallmark, and PepsiCo."[20] Wal-Mart's size and corporate power, of course, render it exceptional here, as elsewhere; but the trend is much broader. Recording companies, for example, have begun to promote bands and albums not only via radio stations and music television but also by piggybacking on what had previously been unrelated distribution systems. Song, Delta Airlines' low-cost unit, exposed various recording-company acts "to a captive audience of air travelers" aboard its fleet of Boeing 757s.[21] To the consternation of music retailers, the singer Alanis Morisette delayed the release through record stores of her new CD by making it available exclusively at units of the Starbucks coffee chain.[22]

Competition from new media, the growth of communications capabilities within nonmedia companies, and the tightly focusing need of consumer products companies to dispose of swelling surpluses have combined to confer on sponsors an additional edge in their dealings, as customer-patrons, with advertising agencies (which shed seventy thousand jobs, 14.4 percent of the industry total, between 2000 and 2004[23]) and corporate-commercial media. The shifting fortunes of members of the institutional triad signal that power relations among them are also changing. As an astute analyst concluded: "The world of advertising turns upside down when the advertisers—not the agencies—are the ones pushing the envelope. But that is what has been happening."[24]

In this context, regardless of how valuable its contribution turns out to be in actually selling commodities, each successive advertising campaign opens a flank in an ongoing war of position waged by the sponsor system across the terrain of daily life.

The Sponsor System and the Quotidian

The sales effort is undergoing a metamorphosis. Geographically, it has attained a worldwide dominion, one from which no economy or culture is exempt. Institutionally, it has metastasized, as new realms of emerging, or formerly independent, cultural practice are successively tied-in or annexed. Gaining controlled access to the quotidian, finally, requires not only the establishment of novel selling venues but also a refreshed and expanded ability to exploit newly available sources of strategic market intelligence. Each of these three avenues of transformation requires analysis.

Relentless commercialization of national broadcast systems, beginning with the United States itself during the 1920s and carrying on to Latin America in the interwar period and then to Britain in the 1950s,[25] gained irresistible momentum during the 1980s.[26] Campaigns waged by advertisers and their emissaries to supplant national public-service broadcast monopolies with corporate systems financed by advertising succeeded in establishing a new worldwide norm.

Multinational synchronization of marketing followed. The emerging global corporate media system pivoted around the measurement and delivery of what advertisers call "most-needed audiences" to the consumer-product marketers that comprise the commercial media's chief patron. Advertisers leap to experiment with their new freedom to develop and stabilize commercial media access to audiences on a supranational scale. Coca-Cola teamed up with the Disney empire to market juice and milk- and water-based drinks in packages emblazoned with Disney characters to children around the world.[27] The Discovery Channel in 1999 synchronized the broadcast of a program about Cleopatra at 9 PM across every time zone around the world.[28] Viacom tested its ability to use its MTV networks as a global promotional platform by striking an alliance with MGM to market the studio's James Bond film (*The World Is Not Enough*) to a teenage audience through the three hundred million households the network reached in Asia, Europe, Latin America, and the United States.[29] Playing to Asian celebrations of the Lunar New Year, McDonald's introduced its Prosperity Burger through shared TV advertisements to nine countries, from South Korea to Indonesia; even as it did so, the fast-food chain "united all of its markets under the 'I'm Lovin' It' theme" through a TV ad shown around the world.[30] Such milestones were inconceivable in any previous era; by the 1990s they inspired continuing debates among practitioners over the advantages and disadvantages of "global branding"[31] and whether global, local, or regional advertising is most effective.[32] The structural issue, of course, is not whether commodities can be packaged and sold more successfully through monolithic as opposed to more focused advertising campaigns but that the sponsor system now possesses sufficient reach and coordinating power to decide whether in a given case suppressing or parasitizing difference is likely to be more profitable.

Although advertising spending is still overwhelmingly concentrated in the rich triad of North America, Western Europe, and Japan, "growth has been faster in Asia and Latin America, especially since the mid-1980s. . . . [Between 1986 and 1996,] individual countries in these regions have shown spectacular advertising growth: for China more than 1,000%, for Indonesia 600%, for Malaysia and Thailand more than 300%, and for India, the Republic of Korea, and the Philippines more than 200%."[33] In 2000, economically impoverished Vietnam opened its market to foreign advertising agencies, working with local partners; several super-groups quickly took up residency.[34] Fast growth also continues in China, which, with $18.9 billion in ad spending in 2004, is sometimes projected to surpass Japan as the world's second-largest advertising market within a few years.[35]

If the sales effort has been extended spatially, within the wealthy metropoles of global capitalism it has intensified. The slump of the early 2000s hit Japanese advertising, but boosted by Upper House elections and the Olympic Games in Athens, spending on advertising in Japan came back in 2004 to over $55 billion. Enjoying fifteen years of rapid expansion in symbiotic relation to the decline of public-service broadcasting, television-ad spending in Europe alone totaled around $27.8 billion in 1999.[36]

Developments in the United States show how, along a second vector of industry expansion, institutional policy has been reoriented to support the assimilation of new fields of practice by the sales effort.

The case against such enlargement of advertising's role within the wider culture is not finally about abstract aesthetic taste or standards, it is about democracy. The tightly limited proprietary interests of the sponsor system are presented as the general public interest wherever they take hold; in actuality, however, they are an inadequate and illegitimate substitute. The standard against which to judge new advertiser-supported media is equally clear. Each new medium might have been made a basis for enlarging the space of free- dom for individuals and societies to deliberate and democratically choose new paths of economic and cultural development. Instead, battles have been repeatedly fought to prevent the actualization of that freedom. As Armand Mattelart aptly puts it, the sponsor system's governing policy admonition is brutally simple: "No media without advertising."[37]

Think, in the present context, of the Internet. Rather than becoming an inde- pendent, open infrastructure for the exchange of textual and audio-visual infor- mation resources, the Web was rapidly colonized and transformed into a new sales instrument.[38] After setbacks and much-hyped projections, online adver- tising increased 20 percent during 2003 to $7.2 billion and grew to $9.6 billion in 2004.[39] Marketers of junk food, for example, were able to seize upon online games as a fresh means of selling, especially to children. A *Wall Street Journal* reporter explained, "Brand-laden diversions, sometimes called 'advergames,' are emerging as a powerful and inexpensive new ad medium, cropping up on dozens of sites from marketers of cookies, candy, cereal, chips, and soda."[40] More generally, a flood of unsolicited junk email has engulfed the Internet; early in 2004 an estimated 80 percent of global email was spam.[41] By compari- son, in 2004 the U.S. Postal Service shipped about ninety-eight billion pieces of first-class mail, which was declining numerically, and 95.5 billion pieces of the growing class ("standard mail") that includes direct mail.[42] By 2005, eight million Americans were publishing blogs, or online journals, and thirty-two million were reading them. Blogs have also begun to be deployed by corporate marketers as "the latest channel for direct marketing."[43] Ad-supported Internet video has begun to take off, finally, as consumer broadband access widens via broadband cable modems and digital subscriber lines.[44]

Other new media are subjected to the same relentless compulsion. A com- munications consultancy found in a 2004 survey of U.S. cell-phone users over age eighteen that "only 20% had received a text ad via cellphone"; Absolut Vodka was among the companies working to redress this deficiency by spon- soring text messages. General Electric's NBC Universal broadcast eighteen to twenty video news reports designed for cell phones to some Sprint subscrib- ers.[45] Despite disincentives (in the United States) like the receiver-pays policy, the do-not-call list faced by telemarketers, and the absence of a cell-phone-number directory, text messages constitute a billion-dollar advertising market, albeit one in which the United States lags behind Europe and Asia.[46]

Where restrictions remain on current or contemplated marketing practice, they are resisted; where ground has had to be conceded, as with U.S. tobacco advertising, new frontiers have been opened, as with U.S. prescription drug advertising. Between 1996 (the year before the U.S. Food and Drug Administration relaxed TV and radio prescription-drug ad rules) and 2003, the drug industry's direct-to-consumer ad spending increased from $791 million to $3.2 billion. Its sophisticated marketing was multifaceted, spawning free samples and product demonstrations for physicians and relying upon one salesperson and nearly a hundred thousand dollars for every eleven practicing physicians in the United States.[47] Would that teacher-student ratios in schools were furnished resources sufficient to approach this benchmark!

Through innovations over many decades, the commanders of the sponsor system have armed themselves with an arsenal of strategies with which to guard and, if possible, advance their interests.[48] Not only did advertisers, advertising agencies, and the owners of concentrated corporate media collaborate in the use of lawsuits, market pressure, lobbying, and public relations to preempt the proliferating channels of public expression; they also presented any and all restrictions on selling as subversive assaults on the democratic right of free speech. Attempting to deflect advocates of government intervention to restrict advertising of sugar- and fat-heavy foods to children, for example, General Mills, Kellogg, and Kraft Foods—the top three advertisers of packaged foods to children, with combined yearly spending on kids' ads of nearly $380 million in the United States—formed the Alliance for American Advertising with other industry organizations "to defend the industry's First Amendment rights to advertise to children."[49] As it is transformed into a corporate prerogative, freedom of speech—the preeminent prerequisite for democratic self-government—is systemically degraded.[50] My local newspaper editorialized, for example, in favor of augmenting still further the First Amendment protections enjoyed by Nike and other corporations: "[C]ommercial speech serves a clear public good because it generates economic activity through the distribution of accurate information to consumers."[51] This would be merely insipid if it did not signal a profound encroachment on democratic procedures.

In the United States, global advertising's heartland, relentless sales pressure, TV ad "clutter," elusive consumers, and a cascade of new media have prompted marketers incessantly to seek novel ways of differentiating their sales messages, and for sponsors, product placement holds growing attraction. One survey found that songs that made it onto the *Billboard* Top 20 chart during the first half of 2004 mentioned fifty-nine different brands a total of 645 times; Hennessy and Cadillac led the pack.[52] Sponsorship saturates daily life so thoroughly that brand references in songs are not necessarily bought-and-paid-for by advertisers. But television, where costs faced by program producers are much higher, is another story. Nine sponsors of the hit show "Survivor" (on Viacom's CBS network) purchased most of the available advertising time in advance and received product-placement opportunities within the text of the episodes shown during its fourteen-week run.[53] A women's talk show hosted by a long-time newscaster, Barbara Walters, "pa[id] homage to Campbell's soup" live on screen, for which Campbell's paid Disney's ABC network an undisclosed sum. In this instance, publicity generated a negative reaction, but the general practice continues to escalate: "In a creeping trend on talk shows and other nonfiction TV formats," noted a reporter, "marketers are finding that their ad dollars have the ability to stretch right through commercial breaks into the heart of scheduled airtime."[54] The same trend is evident in television drama. On daytime television, Butterball turkeys, Nascar shirts, and Kleenex tissue are assigned starring roles. Because advertisers "are increasingly insistent that characters discuss their products," network programmers have told writers to incorporate such references into their scripts. "Writers and producers," reported the *Wall Street Journal,* "often spend hours on the phone with advertisers to learn the nitty-gritty about product-marketing plans, and some shows give advertisers plot outlines."[55] Miller Brewing Company was able to obtain product placement for its beer brands in prime time on the FX channel's firefighter series, "Rescue Me." The media-buying unit of Miller's advertising agency, Publicis, acted as a creative partner with the show's writers, reviewing "scripts in advance to plot out how and where the beer brand might be used." It is not reassuring to learn that, as a result, Miller's beer brands are "excluded from scenes involving talk about alcoholism."[56]

Product placement is not an occasional or narrowly demarcated practice. According to A.C. Nielsen data, the top ten U.S. programs used for product placements during the first quarter of 2005 boasted 12,867 of them, more than half the 23,526 for the top ten shows during all of 2004. The CBS executive Leslie Moonves informed investors that more are coming: "'I think you're going to see a quantum leap in the number of products integrated into your television shows.'"[57] Reaching into a different venue, Publicis arranged for marketers including H. J. Heinz, Procter and Gamble, Kellogg, and McDonald's to pay to have their products appear before and after intermission at a one-time charity performance in London of the musical *Saturday Night Fever*.[58] "While entertainment and advertising are getting cosier everywhere," reported the *Wall Street Journal*, "in Asia they have virtually merged." Philippine clubgoers danced to "Hello Moto," a song commissioned by Motorola; Wang-Leehom, a Taiwanese-American singer, offers what he calls "McHip-hop" in a Chinese rap song called "I'm Lovin' It," the recording of which was originally paid for by McDonald's.[59] In still another illustration, Hollywood has become increasingly focused on profitable tie-ins, whereby animated film characters are licensed to appear in advertisements.[60]

Attempts to co-opt governmental and public-service institutions are especially striking. By 2004 it was increasingly common for university professors with the gift of gab and camera-friendly hair to act as paid shills on TV for corporations and trade associations.[61] Even news broadcasts such as NBC's "Today" show, the top U.S. morning news program, include product plugs by hired experts.[62] More than half of a survey sample of California's high-school districts, meanwhile, allowed sale on campus to students of branded fast-food items including hamburgers, fries, and pizza from franchised chains.[63] While U.S. students' math competence trends dismally downward, primary-school mathematics textbooks sometimes carry distracting references to brand-name consumer products, from Nike to Disneyland.[64]

The sponsor system has claimed still other domains; for example, it is routinely on exhibit at museums.[65] And, in a new-and-improved version of papal indulgences, the Vatican Library permitted corporate endorsements, for a fee, of its Jubilee celebrations; Gruppo Telecom Italia coughed up $80 million worth of telephone and Internet services in exchange for exclusive rights to display its logo on shirts, hats, and umbrellas worn by seventy thousand volunteers at the festivities.[66] The U.S. Postal Service sells advertising space on delivery trucks, collection boxes, priority envelopes, and inside post-office lobbies; stamps, onetime symbols of national sovereignty, are adorned with images of Disney's *Lion King* and other commercial properties, while Stamps.com sells print-it-yourself postage online to corporations.[67] U.S. elections are also remunerative underwriters of corporate-commercial media; in one estimate, political advertising revenues in 2004 constituted over 10 percent of commercial broadcasting revenue, up from 3 percent as recently as 1992. Candidates, political parties, and independent groups spent at least $1.6 billion in 2004 on television commercials (for offices at all levels of government), a hefty increase over the last electoral cycle.[68] Although, according to a business writer, "tradition and taste dictate that the president refrain from product plugs . . . there's no law in the U.S. against presidential endorsements"—and marketers try regularly to obtain them.[69]

Sports, long a treasured vehicle for corporate selling, continue to unfold additional opportunities. Soccer teams throughout Europe held private meetings with U.S. advisers about how to raise corporate sponsorship cash to build new stadiums; by 2000, ninety privately funded and corporate-branded stadiums or arenas existed in the United States.[70] The trend could give rise to incongruity, even incomprehensibility; consider the "Portland General Electric/Solv Starlight Parade Presented by Southwest Airlines," which featured an auto race, "Texaco/Havoline Presents the Budweiser/GI Joe's 200, a FedEx Championship Series Event." In one estimate, during the mid 1980s, fewer than a quarter of U.S. festivals and events had some corporate involvement; by 1999 it was up to 85 percent.[71] Meanwhile, the right to air sports programming on television remains key to commercial success, as Viacom and News Corporation demonstrated in 2004 when they again agreed to pay a fabulous sum—$11.5 billion between them—to retain TV rights through 2010 for CBS, Fox, and DirecTV to National Football League games.[72]

The consequences of annexation extend even further. The quality of information is systematically degraded as the sponsor system redirects the social purpose of a given institution or medium. Across a large and growing range of practice, this requires that cultural creativity—cultural labor—be co-opted and reorganized on behalf of the selling mission.

In the United States, drug companies have turned with alacrity to the Web and made "educational" grants of up to three hundred thousand dollars to support increasingly widely used health sites to showcase their products. As a result, for example, "On a breast-cancer site by HealthTalk Interactive Inc. . . . a panel discussion of new therapies features a lengthy section on the drug Hercepton, made by the site's sponsor Genentech Inc." It is not reassuring to learn that the *Journal of the American Medical Association* ran an HIV/AIDS site funded by the AIDS-drug maker Glaxo Wellcome.[73] The *New England Journal of Medicine,* perhaps the world's most renowned medical journal, apologized in 2000 after it was revealed that it had published nineteen articles that violated its policy barring drug assessments written by doctors linked financially with pharmaceutical companies.[74] Analogous practices resulted in what a *Business Week* columnist called "the corruption of TV health news," corruption as deliberate as it is general.[75] By threatening to withdraw its advertising, Johnson and Johnson, one of the largest U.S.

pharmaceutical advertisers, successfully pressured a major cable network (USA) to cancel the production of a movie it deemed objectionable.[76]

By 2005, the same tendencies had only grown starker. To "neutralize" dozens of influential but "problem" physicians who had not been won over to its prescription painkiller Vioxx, internal company documents revealed that Merck officials planned to dangle before them such incentives as paid-for clinical trials, consultancies, and grants. Though Merck insisted that these were "educational" expenditures, it referred in the same documents to "Expected Outcome/Return on Investment."[77] Articles reporting on the safety and efficacy of pharmaceuticals published in prestigious medical journals are now routinely deployed as marketing tools by major drug companies. Findings in clinical trials are reported incompletely, and results that favor a sponsoring-company's drug are cherry-picked.[78] If they seem to lack commercial promise, treatments that might deliver considerable benefit are buried, as one news report explained: "[I]s anybody telling depressed patients about exercise? Unlike depression medication—which has behind it the research and marketing clout of the pharmaceutical industry—exercise has behind it little research or advertising money. Consequently, the studies supporting it are typically too small to win publication in major medical journals. Without any marketing push, the studies are getting little attention."[79]

Drug companies are only the tip of the iceberg. According to one study, no less than one-third of surveyed U.S. local TV news directors reported "being pressured to kill negative stories or do positive ones about advertisers."[80] In response to what it declared to be "factual errors and misrepresentations," but which instead may have been vigorous pro-consumer reporting, General Motors withdrew its advertising from the Tribune Company's *Los Angeles Times*—expenditures that had tallied $21 million in 2004.[81] Fearing that major advertisers are bypassing print media for the Internet, the newspaper and magazine industries have organized multimillion-dollar three-year advertising campaigns to "reposition" their brands in the eyes of their commercial patrons.[82] The time for asking whether a democratic polity can afford such media bondage is long past.

Ignoring or sidestepping opposition, the apparatus of sponsorship has once again deepened its hold over daily life by trespassing over obstructive social boundaries—of class, race, gender, geography, and political ideology—to tap previously inaccessible or forbidden realms of cultural practice. Victoria's Secret sends catalog shoppers images of young women dressed in underwear redolent of harlotry; Nike pays graffiti artists, who, hoping to stay one step ahead of the cops, once worked with spray paint to adorn subway cars and tenement walls, to translate their art into "hip, retail-ready iconography" for the purpose of selling sneakers.[83] Images of Che Guevara—a hero of the Cuban revolution, killed by counter-insurgency forces in Bolivia four decades ago—have become a popular marketing tool, used to retail t-shirts, posters, books, beanie hats, and children's clothing.[84] A pregnant woman in Pennsylvania auctioned on Ebay the right to advertise on her future son's clothing.[85] On the wholesale level, the New York advertising agency Ogilvy and Mather (a subsidiary of Britain's WPP Group) entered into a marketing joint venture, "Red Force," with the Communist Youth League of China; university students, required to attend training sessions run by the seventy-million-member Youth League, were taught "how to sell Chinese mothers on a new flavor of Tang."[86] Israeli Kibbutzim, communal farming communities dating back many decades and expressing utopian socialist aspirations, have lost their state subsidies; at Kibbutz Gan Shmuel, "help arrived in the form of the Golden Arches," as McDonald's opened a restaurant.[87] A space company proposed launching "a group of large billboards into low Earth orbit" to make them appear as large as the moon to millions of people around the planet.[88] Other marketers have dared to fantasize for the record about a world in which "advertisers will 'narrowcast' messages directly into consumers' brains."[89]

At the institutional level, the sponsor system is already playing a profound role in the process of cultural germination. In the video-game field, for example, "marketing decision-making and game design collapse upon one another," as one careful scholarly study underscored, and "marketing considerations work their way back into the game development process, leading to the creation of games that are from their inception conceived as franchises whose marketing potential can be extended into multiple cultural spaces and constantly renewed over time."[90]

A third trend—an enriched flow of data-traffic in the opposite direction, from individual persons directly to advertisers—not only attests to the lurching dynamism of the sponsor system but also constitutes a disturbing extension of past practice.

Capital's secularly expanding need to market goods and services has for decades inspired attempts to learn systematically about consumers. Within this context, the sponsor system's growing dependence on the co-optation of vernacular practices has likewise prompted sustained efforts to probe the environing culture. However, to touch the pulse of vernacular culture—to make music, clothing styles, popular expressions, and even slang grist for the advertisers' mill—requires unabating surveillance. Spying on the population is also needed, to gauge and predict buying behavior. In establishing advertising rates, advertisers map the demographics of commercial media audiences and chart their patterns of attention to particular programs, genres, and media. All these forms of watching the audience have long constituted a component of the commercial media system.[91] "As public and private national surveys of consumer expenditure multiplied into the hundreds by the early 1930s," writes the historian Victoria de Grazia, "Americans exposed their every nook and cranny to probes and tests."[92]

However, new exigencies have presented themselves. In an effort to define and sustain private-property rights in culture, in a period during which the apparatus of selling is undergoing technological modernization and when the sales effort has to be pursued through the travails caused by surplus production, commercial surveillance has grown increasingly invasive. In one account, "The only way in which corporate owners of copyrighted products can regulate ... possible infringements is to monitor the entire terrain of media consumption."[93] While older data-collection systems and metrics are accorded new critical scrutiny, therefore, measuring practices have plumbed deeper and more generally into daily life.[94] In 1999, Motorola garnered publicity for hiring anthropologists to learn how to sell wireless telephones in Azerbaijan, Kazakhstan, and Uzbekistan (in one estimate, some 2,200 anthropology Ph.Ds worked in "applied" positions in industry, compared with about nine thousand in U.S. academia).[95] Anthropologists, declared one business journal, "once sat in jungles recording the sexual mores of Polynesians but today they are more likely to be in the workplace studying how staff spend their time at the computer or watching ordinary families do the ironing."[96] Audience behavior and shopping patterns are now omnipresently tracked by data-hungry sellers or their agents. TV set-top boxes and digital video recorders, Web-site monitors, databases, and other high-tech systems comprise components of an enveloping eavesdropping apparatus that is well on its way to becoming a virtual sheath over social interaction.

ShopperTrak RCT, a consumer research company, watches consumers across the United States through forty thousand video cameras placed in stores and malls. The company merges the data it gleans from these cameras with detailed information about consumer spending, which it obtains from credit card companies and banks.[97] Wal-Mart TV, an in-store network for advertising to shoppers as they approach the point of purchase, reaches 133 million people a month; according to a WPP executive, in-store ambient TV trials were being conducted by retailers at eight thousand locations.[98] Ogilvy and Mather created a unit to produce video ethnographies of people in natural situations to help advertisers develop better marketing ideas; the subsidiary, Field Brand Investigations, generated a body of work that by 2005 included more than 250 documentaries, "depicting everyone from Thai auto enthusiasts to English schizophrenics."[99] These were adaptations of what quickly became an encompassing industry of video surveillance—an industry that, after 11 September 2001, doubled its growth rate to between 15 and 20 percent annually, and which especially in cities could be relied upon to yield multiple recorded images of crime scenes, suspects, and witnesses.[100]

For ten thousand dollars a year, clients—the Department of Defense among them—can obtain monthly reports on the use of the Internet by some twelve million U.S. primary and secondary school students whose Web-surfing habits are tracked by a private filtering software company.[101] In large part to gain access to sensitive data about users' personal health, Omnicom, an ad agency super-group that represents major drug companies, has taken financial stakes in five major healthcare-related Web companies.[102] Rampant use of "cookies" (online tracking devices that generate great quantities of rich customer data about Web-surfing habits) has even spawned fights over control of the resulting data between big corporate marketers and their own Internet advertising networks.[103] For corporate planners, tapping into the new data generated by Internet service providers and other online services holds strategic allure: "AOL is probably sitting on a bigger wealth of information about consumers than any other entity," enthused one advertising-agency executive when AOL led the pack of Internet service providers.[104] In the runup to AOL's takeover of Time Warner, business analysts expressed delight at the prospect of gaining access to immense amounts of subscriber information, as the combined company pursued "relationships" with 130 million subscribers to its magazines, cable television systems, pay television units, and Internet-access services.[105] Similar hopes quickly attached to the wireless technology called radio frequency identification, which permits tiny chips attached to individual commodities to track their use.

Conclusion

The scale on which advertising can be normalized as a cultural project continues to exceed expectations. When Coca-Cola donated a half-century's worth of its TV commercials from around the globe to the Library of Congress, these turned out to number more than twenty thousand—meaning that each day for fifty years, on average, at least one new Coke advertisement was rolled out somewhere.[106] What other social actor besides the modern transnational corporation could claim such a ubiquitous presence? What other cultural practice was allocated comparable resources?

The sponsor system only continues to grow, and its expansion is to be measured not simply by the saturation of public and private life by manipulative words and imagery. Regardless of whether manipulation succeeds in any given instance, an overarching shift in social priority is now far advanced. It goes beyond the fact that newspapers set up sections on style, business, and sports to bundle readers into sponsor-friendly packages, and that television networks rely on invariant series formats in hopes of predictably reaching—and selling—audiences to advertisers, and that magazines and Internet sites stir and blend editorial and advertising matter at the behest of sponsors. It goes beyond the fact that, on behalf of advertisers, genres are preferred for reaching "most-desired" demographic groups while excluding others, and that the fate of individual programs and titles—indeed, of entire media industry segments—is decided by their value to commercial sponsors. The circuits of daily life are being set at the service of the sales function so that virtually any area of cultural practice can be reorganized to suit the demands of its underwriters. Captured by proprietary interests, the culture skids and slides away from democratic development. Mobile communications add additional distinctive contributions to the tyranny of market compulsions.

8

Mobilized

"Cutting the cord" unleashes a giddy new freedom of communication: via wireless devices we may speak with the world even as we move through it, giving rise to a condition that some analysts call "perpetual contact."[1] Habits of dependence, already widespread and well entrenched, are rapidly diversifying. During 2003, over five hundred million mobile handsets were sold worldwide; as mobile handsets in use came to outnumber landline telephones, wireless bulked ever larger within the overall culture and within the telecommunications sector. One-third of Japan's population, some forty million people, used its top carrier's mobile Internet service; U.S. subscribers spent 912 billion minutes on mobiles; Europeans sent 113 billion text messages via short-message service, a shade over half the number of text messages exchanged in China.[2]

It is at least vaguely obvious that the need felt by individuals for wireless communication is of social origin. It is less obvious that, through both direct and contradictory forms of pressure, this need has been shaped by those who control the means of production and who are interested in extending the trend toward individualism.

Supply-Side Pressure

In the aftermath of the Internet bust, new traces of supply-side pressure could be glimpsed. A huge new wave of capital, incessantly searching for new and profitable investment outlets, fastened on mobile telecommunications.[3] The offerings of informationalized capitalism's omnipresent publicity apparatus swiftly accomodated the new speculative emphasis. Newspapers, magazines, television, and Internet sites reiterated, in seemingly effortless synchrony, the theme that wireless constituted the next new thing. The boom that would materialize around wireless, opined L. John Doerr, a veteran venture capitalist who learned his craft by backing Netscape and Amazon during the go-go years of the Internet bubble, would be "'ultimately more important and will likely offer a larger wave of investment opportunity than the personal computer.'"[4]

Behind the hoopla, corporate capital was mobilizing to make the most of whatever opportunities might be presented. QualComm, Motorola, Intel, Nokia, Sony Ericsson, Samsung, Vodafone, Hutchison Whampoa, NTT DoCoMo, Microsoft, and other high-tech corporations have placed stupendous bets on competing wireless gizmos and services. Within every identifiable niche of the growing mobile economy, market-development efforts have taken hold, from specialized microelectronics and operating-system software to network services and new applications. In 2005, no less than twenty companies were vying to sell high-speed wireless equipment to the U.S. residential market alone.[5] With 170,000 cell-phone towers jutting into U.S. landscapes—evidently still an insufficient number to ensure adequate service—independent investors have built additional towers on speculation, hoping for profitable sales to one or another of the carriers.[6] Wireless incursions into daily life are duly proliferating.

The publicity apparatus pounces upon fresh evidence that wireless markets are being successfully cultivated, to test it for signs that high-tech's former exuberance might be resurrected. During 2004, U.S. subscribers purchased $4 billion worth of wireless data services (chiefly to send text messages).[7] Might this signify that risky, multibillion dollar rollouts of "third-generation" networks to furnish high-speed data services will pay off? It has been prognosticated that cell phones in use worldwide "could rise 40% to more than 2 billion in three years." Is this a portent of continued stable growth for carriers?[8] Wi-Fi and WiMAX, new wireless technologies promoted by Nokia and Intel and other major vendors for use with notebook computers and other handheld devices,[9] offer another conundrum: which is likely to win out among consumers? People carry cell phones more than any other device, according to one U.S. survey. Is the destiny of wireless to steal the market for mobile music away from Apple's iPod?[10] (As this book went to press, it did not seem so.) Two of the world's largest recording companies, Universal Music and Sony Music, are asking star musicians to remix songs into ninety-second versions designed for sales through mobiles.[11] What about TV shows, video games, other types of "content"?[12] How might their financial prospects and business models be affected by the tools of electronic mobility?

Once again, we are witnessing a breathless promotional effort and a debilitating social regression. The marvel of mobile connectivity, another product of corporate-led reconstruction, has incarnated predatory and chaotic tendencies.

Incompatible networking standards developed by rival corporate consortia, by locking in subscribers, have balkanized world wireless markets, making it more difficult for travelers to make calls or access data from a single device. Competing approaches to next-generation mobile technology stand to multiply these difficulties.[13] In the United States, the consolidation of the wireless phone sector is already compounding the problem of incompatible technical standards, as Cingular, the nation's largest carrier, tries "to push former AT&T wireless customers to move from their old TDMA network to Cingular's newer GSM network, which requires those customers to upgrade their phones."[14]

Rapid obsolescence is hardwired into the mobile phone market at several levels. During the 2004 season, sharp color-screen clamshell designs vied with candy-bar-style models and slide phones; fads included a push-to-talk feature, built-in cameras, and faster speeds. "Top players such as Sprint PCS Group and Verizon Wireless in the U.S., Vodafone Group PLC in the U.K., and Hutchison Whampoa in Hong Kong demand the best or hottest product that's available anywhere in the world. They stand to gain because more-sophisticated phones can be used to access the new services that operators are heavily marketing to customers these days. That increases traffic and profits for the network operator."[15] Competing phone manufacturers thus pay a heavy price for missing the latest fashion, and large vendors such as Motorola habitually launch fifty to sixty handset models each year.[16] But this profligacy exacts an even steeper price from society at large; as their life span contracts (down to around eighteen months by 2004), toxin-filled cell phones become disposable items. With no regulations mandating recycling, in the United States alone subscribers are trashing at least twenty-five million handsets annually.[17]

East Asia, significantly, functions as global mobility's pacemaker, at least in the consumer market. In Japan, fierce rivalry between the industry leader DoCoMo and the upstart KDDI helps propel ceaseless experimentation.[18] During 2003–4, 20 percent of DoCoMo's revenue—about $9 billion—was generated from the transmission or downloading of data by forty-two million users of its "I-mode" Internet service; by 2005, it was garnering nearly 26 percent of its revenues from data services. What types of content drove this growth? Horoscopes, songs, games, ring tones, movie clips, and email.[19]

But the more basic spur to enlarge and extend wireless service markets worldwide has come from what is, for the telecommunications industry, a wearisomely familiar source: slowing growth. The existing market for wireless voice calling is growing increasingly saturated.[20] As the journalist David Pringle commented aptly in 2005, "While handset sales are booming thanks to the addition of cameras, music players, and fancy software, cellphone voice services are fast becoming a basic commodity distinguished primarily by price."[21]

In chapter 5, I traced key features of the crisis in telecommunications; wireless-system development shares some key traits with its wireline cousin. A familiar progression has etched itself into the wireless field: rivalrous investments by carriers in the most tempting national markets has led first to rapid expansion, then to a slowdown in revenue growth, and then to rate-cutting, consolidation and an increasingly intense profit-scramble.

As saturation develops in Western Europe and Japan, which together make up nearly half of the global market for cell-phone service, wireless companies are racing to develop new services. Operators were already pressing into multimedia applications early in the 2000s.[22] With a further wave of network investments by a diversifying group of service providers and manufacturer rollouts of devices designed for multimedia data exchange, corporate experimentation to learn what would sell had grown more intense and pervasive by 2004.[23] One old standby, a staple of each new medium from printing onward, again demonstrated potency; by 2005, worldwide spending on video downloads of pornography to over fifty million mobile phones equipped to receive it was projected to top $1 billion.[24] Other vices also evinced promise. In 2005, Nevada was on the verge of permitting gamblers at resorts to play video poker, blackjack, and other games on personal digital assistants and other handheld devices.[25] Industry executives, explained David Pringle, are well aware that "the industry must find new hit services to go alongside voice- and text-messaging in order to persuade people to spend a greater chunk of their income on cellphone services."[26]

The ferocity of the economic forces driving this later phase of system development is confirmed by another, seemingly disparate, trend. Hoping again to obtain growth, major equipment vendors and carriers have turned increasingly to less well-endowed regions. Motorola and Nokia are building low-end handsets for sale in "emerging markets" such as China and India, where growth rates continue to be higher; the same motivation has impelled service providers such as Vodafone to acquire systems in Eastern Europe, China, and even Africa.[27] This is usually presented in quite different terms, of course. Africa, for example, found itself heralded in 2005 for having become the world's fastest-growing mobile-phone market, and carriers were patted on the back for delivering the continent—albeit on a prepay model for customers concentrated in only four countries (South Africa, Morocco, Nigeria, and Egypt)—from having to route calls through former European colonial capitals.[28]

No guarantees can be offered that this strategy will succeed. Midway through 2005, third-generation systems and multimedia applications appeared to be catching on in some places. Far from making good the shortfall that is emerging in voice-service markets, however, the new offerings appear equally likely to cannibalize their forebears, as subscribers substitute email and text messages for voice calls.[29]

Wireless-system development therefore remains fundamentally unpredictable and contingent. Its spectrum-dependence—its need for access to huge chunks of the electromagnetic spectrum, the invisible resource on which free-to-air broadcasting, satellites, and other forms of electronic communication rely—has already engendered other market instabilities. During 2000 and 2001, at the height of the high-tech bubble, panicky wireless carriers ponied up huge sums in government spectrum auctions—$33 billion in the United Kingdom, $48 billion in Germany, $17 billion in the United States—to guarantee that they, rather than rivals, would be enabled to build out next-generation wireless networks. The debt many of these companies assumed to do this, coupled with astronomically priced buyouts of competitors (notably, Vodafone's $181 billion hostile takeover of Mannesmann in 2000), helped trigger the telecom industry meltdown of 2001–2.

In the aftermath of that debacle, tens of thousands of jobs were eliminated, disproportionately based in the more heavily unionized telecommunications-manufacturing and wireline-service segments. And owing to rival network expansion, overcapacity continues to burden the field. Ruinous competition within a glutted market carries other negative consequences, in addition to mobile operators' sharply reduced ability to set prices.

In the United States, complaints about poor service quality and questionable billing practices have persisted.[30] Hidden fees and surcharges disguised as government-imposed taxes are tacked onto bills by major U.S. carriers, so that 11 percent of respondents to one 2003 survey said they had experienced serious or continuing billing problems.[31] Another study reported that no less than three-fifths of all U.S. wireless subscribers called customer-service centers at least once during 2002 to complain.[32] A U.S. government report described lack of coverage, static, network overloading due to insufficient capacity, and dropped calls as frequent sources of complaint; one-fifth of U.S. users were unable to complete 10 percent or more of their calls.[33] A survey of 21,700 wireless customers by J. D. Power in 2004 showed no change from the previous year: around one of three cell-phone calls suffered quality problems of one kind or another.[34]

Is this merely the price of progress? A *New York Times* correspondent claimed that "Americans' use of cellphones has increased so quickly that wireless networks are becoming overloaded, causing a growing number of customers to complain."[35] But this is only a secondary factor. Service problems actually stem from the fact that, in the United States, as befits the historical source and global center of neoliberal reconstruction, "wireless phones [are] much less reliable than their wired ancestors."[36] Even during emergencies (wireless companies make safety a selling point when advertising their services), mobile-phone networks' performance is often inferior to that of landline systems.[37] Though often derided by well-placed devotees of the new new, the old landline network was engineered to reliability standards that exceed those used to build out wireless systems and fiber-optic cable systems.[38] In the United States, at least, it is difficult not to conclude that the younger generation is being trained to accept lower-quality telecommunications service.[39]

The institutional basis of the wireless industry is to blame, not its growing pains. U.S. wireless networks typically lack adequate backup power generators, unlike the earlier landline telephone system, which carried its own power along with conversations. Moreover, U.S. wireless networks have been typically underbuilt; they are not constructed to handle great surges in call volume, as was demonstrated when the attacks on 11 September 2001 overwhelmed them.[40]

Both flaws are remediable. By establishing backup power sources and improving network coverage and capacity, mobile-service providers could come closer to approximating the renowned "five nines" reliability of older U.S. landline networks, which afforded call completion 99.999 percent of the time. But this would require large increases in capital expenditures. And, as one industry consultant explained, "A service provider can't do that and stay in business."[41] Here we arrive at the underlying problem. The competitive market beholden to neoliberal policy makers has engendered overcapacity across the greater telecommunications industry and—paradoxically, but predictably—inadequate network investment by individual carriers.

Government overseers offer feeble assurance of relief, as the wireless industry remains much more lightly regulated than its wireline forebear.[42] Although the Federal Communications Commission possesses jurisdiction, it is characteristic that, after fifteen years or so of explosive growth for mobiles, as of 2003 the FCC hadn't even established a minimum level of wireless call quality.[43] Nor are there federal regulations requiring carriers to maintain a certain amount of backup power. The comparatively poor reliability of wireless networks is a harbinger of a wider problem, because wireline carriers, as they have invested in fiber-optic networks, have also begun to rely on commercial power providers rather than on their own electricity.[44]

The industry sails on, but it moves atop a thin skin of ice that places users in danger of falling through. For all the hue and cry about homeland security, emergency service for mobile-phone callers trying to reach help through 911 remains woefully inadequate. Though cell-phone callers originated more than one-third of the 190 million 911 calls in 2005, only two-fifths of the country's six thousand emergency call centers could locate the source of these calls. This is not a mere developmental lag; in 1996, the FCC asked the increasingly fragmented industry to devise a remedy, calling on carriers to install upgraded emergency service on a comprehensive nationwide basis. However, according to Anthony Haynes, the executive director of the Tennessee Emergency Communications Board, wireless companies successfully resisted, falling back on the characteristic neoliberal argument, "'We can't do this, our industry is in its infancy, and these costs will stifle growth.'"[45]

A different factor helps explain why there exists no directory of wireless subscribers: too many subscribers don't want one. Users of the nation's 182 million cell phones are thus forced back on their own resources when they want to obtain telephone numbers they do not already possess. Citing privacy concerns, consumers have said that they prefer not to be available to callers apart from those to whom they have given their number privately—not least because they usually pay for incoming calls. An additional disincentive springs from consumers' experience with the National Do Not Call Registry, established to rein in telemarketers using the wireline network. After two years, the registry had enrolled over a hundred million numbers, meaning that with some exceptions, telemarketers were supposed to cease calling them. However, half of registered consumers said they continued to receive calls they think the list should block, and they lodged fully one million complaints with the Federal Communication Commission.[46] Consumer resistance and wariness thus have helped stymie development of an everyday shared resource of obvious utility. "'People would love to be able to contact each other,'" explains the communications scholar James E. Katz, "'But they are very reluctant to be reached.'"[47] Perhaps no other indicator so reveals the social wreckage on which the marvel of mobile communications is being erected.

The New Face of Mobile Privatization

The massing of corporate power does not explain everything. To get at other aspects of the phenomenon, it needs to be underlined that the global love affair with wireless signifies a volcanic release of a pent-up demand. Whence this urgency? Why, over merely fifteen years, have wireless handsets come to outnumber landline telephones?

Some have suggested that surging demand arose as a reflex of the comparative historical scarcity of landline telecommunications investment, especially but not only throughout the global South. When, beginning in the 1980s, capital at last commenced to modernize world telecommunications, demand therefore naturally eclipsed any prior estimate. True enough. But why did this demand channel so impressively into wireless? Wherein lay the need for *mobile* connectivity?

This need, again, is *socially* established; there is no innate human preference for "perpetual contact." Historically changing social relations comprise the crucible in which specific technological potentials are forged or allowed to fail.

The need for constant connectivity, in all its multiformity, signifies a transition into a new phase of "mobile privatization"—a trend identified in 1974 by the great Welsh cultural critic and theorist Raymond Williams. Williams gave this name to a complex of interweaving processes involving movement away from smaller-scale production and settlement and the rise of the family home distanced from places of work and government. Mobile privatization is no mere mechanical outcome but a complex and contradictory result of social struggles; necessity and desire have no obligation to separate themselves out in neatly opposed arrays.[48]

Urban and suburban sprawl is the built environment engendered by mobile privatization. To make such a landscape, alongside automobiles and shopping malls and endless parking lots, new forms of communication are vital. Williams employed the idea of mobile privatization to analyze the development of broadcasting, which provides a means of keying millions of individual households into central sources of shared news and entertainment.

The wireless phenomenon constitutes a wrenching extension of this deep-rooted historical tendency. Wireless accomplishes this in part by building upon foundations laid by its landline ancestor. At least within developed market economies, landline telephones offered a key element in living arrangements built around private households that could remain "in touch" with surrounding communities. More than a century ago, the rules of etiquette for telephone usage began to be worked out, and the contours of daily life actively redrawn, to accommodate what appeared to be the useful, pleasurable, and yet also invasive and irritating telephone.[49] This necessitated cultural work beyond any mere carryover of conventional decorum or the crafting of functional codes for telephonic conversation: beginnings, closings, turn-taking cues, and sequences. Deeply felt alterations in the social distances between home and work, between male and female, and between class and class had to be renegotiated to incorporate the new medium.

A comparable process can be glimpsed around wireless. How frequently we hear the complaint that cell-phone users are loudmouths, or that they make the rest of us into involuntary eavesdroppers. But more importantly, on the street, in schools and offices, on the highway, in restaurants, and across the varying milieux within which mobiles are used within culturally distinct national environments, our daily social life is being twisted and reshaped.

Advertisements endlessly enthuse over how mobiles enhance individual freedom by breaking with arbitrary, or at least stuffy, limits. News accounts tend to focus on the other side of the same thing: how wireless is used to subvert conventional expectations. Press portrayals cite school children using mobiles to flout their teachers' admonitions; prison inmates using mobiles to plan felonies; dissemblers of all sorts wirelessly transgressing cultural norms and undercutting prevailing institutional practices.[50] But the pleasures and perturbations of an expanded individual freedom are the medium, not the essential analytical benchmark, for evaluating how demand for mobiles is anchored.

Opportunities for individual and family empowerment motivate innumerable decisions to integrate wireless into daily routines. But this freedom is layered and molded by gross inequalities and other forms of compulsion. When work and leisure are unevenly distributed, when longer working hours are coupled to high levels of unemployment; and when deliberately diminished social-service provision adds measurably to the hours and energy that must be expended on getting through daily life, then, in innumerable ways, for most people the take-up of mobiles constitutes an attempt to solve privately what appear as intractable problems of reality.

Underlying this extension of mobile privatization around wireless is also the changing social experience of wage labor. This is no simple or uniform process; the wage relation contributes in complex and contradictory ways to the social patterning of "perpetual contact." Mobiles enable us to seize time within the workday for personal purposes, even as they also make us newly accessible to the boss. Mobiles allow us to more efficiently exploit what time remains after work. They thus contribute to sensations of freedom even as they concurrently assist incursions and constraints on that same freedom.

Consider wireless development in China, which, with 269 million subscribers in 2003, has become the world's largest wireless market, and where the social need for mobiles is bound up not only with the rapid growth of an urban middle class but also with internal migration on an unprecedented scale. This is widely driven by the newly reorganized need for paid work and by the superexploitation of long working hours. According to the deputy director of the State Administration of Work Safety, during 2004, "Only 30 percent of workers work eight hours a day in the Pearl River Delta. Some 46 percent work fourteen hours a day."[51]

In the United States, by comparison, the workday and the work week are shorter; but working hours have lengthened by one-fifth since 1970, chiefly as a result of falling real wages.[52] The linked trend for women increasingly to enter the paid labor force has ramified, as women continue disproportionately to perform household labor, within larger and larger homes, on a "second shift."[53] Within the United States, wireless use has been widely stimulated and conditioned by these developments, as well as by the decline of social supports—a decline that could not fail to complicate and enlarge the field of responsibilities for dependent care and household management.

That this has occurred atop dispersed landscapes of work and consumption adds further momentum to mobile privatization. Through mobiles, as in the converging area of computer technology, it is often observed that work has been freed to expand into what used to be nonwork time. The automobile is especially crucial here. Americans make 405 million domestic business trips annually to destinations at least fifty miles from home, four-fifths of which are by automobile.[54] Add household shopping, trips to the doctor, errands, social and recreational visits, and commuting, and Americans take 1.1 billion trips a day, 87 percent of which rely on private vehicles.[55] (Walking, in contrast, accounts for less than 9 percent of the total, while public transportation, including school busing, claims just over 3 percent.[56]) Every day, on average, each American driver spends nearly an hour behind the wheel.[57] One study found that U.S. rush-hour drivers spent three times as long stuck in traffic in 2003—forty-seven hours—as they did in 1982.[58] It is therefore not a coincidence that 40 percent of all U.S. wireless use—during

2003, four hundred billion minutes' worth—takes place in the cars steered by our millions of "road warriors."[59] At any given moment in 2004, 1.2 million drivers were talking on a cell phone.[60]

Automobile-based transportation constituted previous generations' enduring "achievement" in furthering mobile privatization. The car was not a technological complement of our social arrangements; in its own right, it became a structuring force. Wireless has built on this irrationality.

Public health has been a casualty of mobile privatization thus reconfigured; think of the highway carnage that now must be attributed to the telephone. In 2002, a Harvard University study estimated that U.S. drivers talking on mobiles were responsible for about 6 percent of auto accidents, killing as many as 2,600 and injuring 330,000 others.[61] More conservative analysts have charged that these figures are inflated and attempted to prove that wireless use does not unsafely distract drivers. But when studies show that drivers with handheld phones have both hands on the wheel less than 1 percent of the time (hands-free headsets increased this figure to 16 percent), it is hard to deny that substantial death and destruction might be attributed to drivers' use of mobiles.[62] Thankfully, numerous countries, three U.S. states (New York, Connecticut, and New Jersey), and many municipalities have banned the use of handhelds by drivers anyway. Government agencies or advisory groups in Britain, France, and Germany have discouraged children from using mobiles because of health concerns about sustained exposure to electromagnetic emissions from handsets.[63]

The frenzied global take-up of wireless signifies nothing other than an attempt to rationalize the irrational by attempting to wrest a measure of personal control in a social world that continues to spin out of control. Wireless does not exist merely on one side of this equation. Means of anchoring mobile connectivity to an increasingly authoritarian capitalism via location-based tracking of individuals and things for marketing and surveillance have already been glimpsed. Mobile commerce, for example, intends to transform the handset into an formidable instrument of the sales effort. Wireless companies like Vodafone hope "to provide location-specific information to a person's mobile device, wherever he may be." Some of this data-stream may be traffic, weather, or news; however, a growing share, if the carriers get their way, will be telemarketing messages.[64] Already by 2005 a California consumer organization relayed customer complaints sparked by their having been billed by their carriers for unwanted advertisements.[65] Australian subscribers to Telstra wireless service are harnessed even more directly to the sales effort: they can call the "Dial-a-Coke" phone number posted on many vending machines, wait for their purchase of a soda to be authorized, and find the charges—for the drink and the call—on their next wireless bill.[66] In Europe, wireless carriers have established a consortium, Simpay, to provide a standard means of using mobiles to pay for goods and services across the continent. A comparable initiative in under way in the United States.[67]

Surveillance practices are likewise being chillingly extended and normalized. Mobiles are praised for allowing parents to put their children on a wireless leash; in South Korea by 2005, 3.6 million subscribers to the "Find Friends" commercial service could track the location of a mobile-phone user to within ten meters.[68] Less has been heard about how police agencies and employers are adapting technology created to track soldiers and prisoners to generate torrents of mobile-phone-based customer-location data.[69] A study of Samsung's own proprietary deployment of wireless technology underlines that the scale of surveillance is expanding from spatially restricted workplaces to the society and the population at large.[70] Consumer goods tagged with radio frequency identification (RFID) technology promise to accomplish a similar feat for commodities. Big users, led by Wal-Mart and the Defense Department, hope to use RFID to gain enhanced control over their vast supply chains; suppliers, including IBM, Microsoft, Oracle, and SAP, look to these "sensor networks" as a new market. "'We are moving from batches of information about operations to continuous visibility,'" explained IBM's manager of its "pervasive computing group."[71]

The world of wireless was not created to deliver us into an era of playfulness and personal freedom. It came to us, rather, as a complex historical extension of the domination and inequality that continue to define our divided societies.

PART 3

Poles of Market Growth, or a Deepening Crisis?

9

Open Questions about China, Information, and the World Economy

Two vectors of change converge on a final, and vital, entry point for analysis. First, despite the bursting of the high-tech financial bubble, a global communications and information industry continues to function as a fountainhead of economic transformation. Second, with its embrace of capitalist social relations, China has become "the fastest-growing large economy in the world" and an engine of market renewal.[1] Are these two poles of growth related? What links are being forged between capitalism's most dynamic sector and its most expansionary growth zone?

Let us begin to explore this issue by reviewing Chinese initiatives in information and communications that have unfolded across a wide array of manufacturing, content, and service markets.

Chinese Initiatives

China's media content and hardware industries and Chinese media advertisers are establishing transnational affiliations and using them to broaden and reorganize the domestic market—and vice versa—at a furious pace. In 2002, Xinhua Financial Network, a Hong Kong joint venture in which China's state-owned Xinhua News Agency claims a minority stake, purchased the Asian business-news operations of Agence France-Presse.[2] In 2001, China allowed AOL Time Warner and News Corporation to transmit Mandarin-language television entertainment channels into Guangdong Province via Chinese Entertainment Television Broadcasting (CETV) and Star TV in exchange for the two media conglomerates' agreement to carry China Central Television

(CCTV) 9's twenty-four-hour, mainly English-language channel over U.S. cable systems.[3] In 2003, however, after falling from its comfortable perch in the aftermath of the Internet bubble, a straitened AOL Time Warner sold a controlling interest in CETV to Tom.com, a Hong Kong-based media group controlled by Li Ka-shing.[4] In 2004, Chinese policy makers lifted the ban on foreign investment in the domestic TV-program industry, allowing joint ventures in which Chinese companies possess a majority stake to be formed. An important additional requirement is that these partnerships create a unique logo to ensure that they did not simply promote foreign media brands within the domestic market.[5]

Chinese production lines, meanwhile, are generating a torrent of television sets. Springboarding from China's domestic market, the world's largest with sales of thirty-three million sets in 2002 (compared with twenty-five million in Europe and twenty-six million in the United States),[6] half a dozen domestic manufacturers have set their sights on exports—which accounted for around fifteen million sets a year—to three dozen countries.[7] In 2003, this export strategy was coupled with foreign direct investment. At the foreign company's initiative, China's second-largest set maker, TCL, originally established by the Chinese government and subsequently privatized, partnered with Thomson (the French electronics manufacturer that owns the RCA brand) giving majority ownership and control of the world's largest transnational TV manufacturer to the Chinese company.[8] With some twenty-nine thousand employees and a dozen factories scattered throughout China, France, Mexico, Poland, Thailand, and Vietnam, TCL-Thompson Electronics was manufacturing around twenty million sets annually by 2004.[9]

Other linkages have been forged to extend and enlarge the reach of the transnational sponsor system. A Chinese company, Yanjing Beer Group, sponsored a U.S. National Basketball Association team in 2003, the Houston Rockets, which, not coincidentally, showcased the the Chinese-born center Yao Ming. About thirty Rockets games were broadcast in China during 2003, so tens of millions of TV viewers saw Yanjing's billboards, which encircled the Houston arena. Yanjing Beer was only one of several Chinese merchandisers to craft sponsorship deals with U.S. and European sports teams.[10] But Yao Ming began to function as a center not only for the Rockets, and Yanjing, but also for the Chinese-U.S. consumer-marketing complex. Working through the National Basketball Association, which by 2004 had crafted deals to broadcast eight games a week into three hundred million Chinese households, Rockets games played in China also drew Eastman Kodak, McDonald's, and Anheuser-Busch as sponsors at nearly $10 million each. Reebok, opening a new store in China every other day in a bid to catch up to Nike, launched Yao Ming's signature shoe in Shanghai in 2004, occasioning a near-riot by his fans, an event likened by an observer to the Beatles' U.S. tour of 1964.[11] However, the undoubted centerpiece of China's burgeoning effort to insert itself into the circuits of a transnational marketing complex will be the 2008 Olympics, which some believe will transform China into the world's second-largest advertising market, surpassing Japan.[12]

The cultural and economic power of this quickly forming sales complex is amplified by a reciprocal development of the consumer payment and credit mechanism. Prior to further liberalization in 2006, after which foreign banks will be permitted to make loans directly to Chinese customers, transnational financial services companies were required to market to consumers through a local partner. Banking giants, including HSBC Holdings, Citigroup, and American Express, duly acquired stakes in Chinese affiliates—the Bank of Communications, Shanghai Pudong Development Bank, and the Industrial and Commercial Bank of China, respectively—so as to enter the domestic market for financial services. By 2005, an estimated twelve million credit cards were circulating, mostly among the fifty million individuals earning more than five thousand dollars a year, alongside eight hundred million debit cards that could not be used to purchase goods on credit. Occasioning rose-colored prognostications concerning the future size and buying power of the country's middle class, the value of bank-card transactions has grown rapidly. There is little question that a formidable new system of consumer debt-finance is being erected.[13]

Comparable trends are apparent in telecommunications, where profit-making opportunities around new technologies have combined with strategic leverage over the domestic market to support entry into transnational markets by expansion-minded Chinese equipment suppliers and system operators. By producing more than twenty-three million handsets built using components from dozens of outsourcing companies, twenty to twenty-five Chinese mobile-phone manufacturers, including TCL, Ningbo Bird, and Haier Group, acquired control over one-fifth of domestic sales by 2002; by the end of 2003, Chinese companies controlled an estimated 40 percent of an enlarged home market.[14] Foreign wireless companies recaptured some of the domestic Chinese handset market during 2004, but what the *Wall Street Journal* called "explosive growth" by Chinese phone companies is "set to overflow onto the world stage."[15] Again, this is part of a larger pattern. In a simultaneous bid to shape domestic-market development and to use this base as a launchpad into the world market, Chinese policy makers are encouraging the development of homegrown technical standards for telecommunications and media products. To the dismay of U.S. corporate rivals and, it is important to add, again with mixed success, this strategy has been adopted not only for mobile phones but also for next-generation wireless networking gear, high-definition television, electronic imaging technologies for the digital home market, and DVDs.[16]

Other initiatives in international telecommunications have attempted to take advantage of important historical contingencies. On the one hand, depressed market conditions continue to predominate throughout this sector as a consequence of the huge (albeit uneven) buildup of network facilities in the 1990s. Major international network assets thus remain on offer at bargain-basement prices. China's own home telecommunications-services market, however, is atypical, even—given its size—unique in that it remains in the hands of domestic capital and the party-state. Averting market entry through the go-go years of the 1990s, when dozens of national telecommunications markets were reorganized by outside interests, China hardly began to allow foreign investors to participate in the development of its national telecommunications infrastructure until after it concluded negotiations to enter the World Trade Organization in 2001. Its major carriers were thus not only able to emerge unscathed from the turmoil that engulfed global telecommunications during 2000–2001; they also continued to be sheltered from competition by foreign carriers because carriers headquartered in the United States, Japan, and Western Europe were intensely preoccupied with paying down their own inflated debts and escaping from ruinous competition in their own home markets and thus no longer in a strong position to invade the Chinese market. Although Chinese authorities have licensed a succession of rival carriers, and competition in domestic mobile and wireline services is intensifying, China's telecommunications operators enjoy a privileged role in developing what has suddenly materialized as the world's largest national telecommunications market (by subscribers). After two decades of unbroken and unprecedented expansion, China's carriers serve an estimated 214 million wireline and perhaps 250 million wireless subscribers (the number of wireless subscribers is difficult to measure because members of well-off social strata frequently pay for multiple services); additionally, there are a hundred million cable TV subscribers.[17] When China's leading network operators train their attention on the exceptional opportunities for transnational expansion, therefore, they do so from what historical circumstances have rendered a relatively strong domestic position.

In this regard, they are not completely unique. As we saw in chapter 4, Mexico's Telmex, which likewise managed to hold on to its dominant domestic-market presence, snapped up distressed network assets throughout Latin America, while thriving Indian telecommunications companies purchased international networks built by U.S. interests.[18] But Chinese carriers' transnational expansion is still noteworthy. A consortium led by China Netcom, the country's second-largest wireline carrier, paid an initial $80 million to acquire a partially owned affiliate of the bankrupt Global Crossing—a unit whose book value was $1.2 billion. After China Netcom bought out its partners early in 2004 to become sole owner, this regional system, renamed Asia Netcom, passed entirely into Chinese hands and began to compete for traffic.[19] Another telling indicator of the carrier's growing transnational orientation came as China Netcom listed its stock in Hong Kong and New York during 2004—a move managed by Goldman Sachs and China International Capital—with a further suggestion that Netcom might deploy the proceeds from its billion-dollar initial public offering of stock to finance additional transnational growth. In this case, however, the depressed state of the telecommunications industry worked against the Chinese company's ambitions, as investors remained skittish.[20] Hutchison Whampoa, based in Hong Kong and controlled by the billionaire Li Ka-shing, acquired mobile-telecom network assets in nine (mostly European) countries and built out third-generation mobile-phone services there in advance of rivals. Hutchison's third-generation services claimed nearly six million subscribers by the end of 2004 and drew on overall capital investment of a reported $25 billion but has thus far achieved only mixed results.[21]

In semiconductors and computing, dynamic interactions between home-market development and the transnational industry are again uppermost. In one estimate, Chinese demand for semiconductors, for products fashioned to satisfy domestic demand and for exported commodities, soared from less than $7 billion in 1997 to more than $28 billion in 2003. (China housed 4 percent of global chip-making capacity in 2003, compared with 18 percent for North America.) To tap into this demand, several Taiwanese-backed startup chip manufacturers set up shop in China, sustained by investments of nearly $10 billion over the past three years, with an additional $5 billion pledged.[22] But, boosted by state subsidies, perhaps as many as five hundred startup chip-design companies have proliferated.[23] China's policy of supporting domestic capital's entry into chip making through tax rebates was withdrawn in response to protests within the World Trade Organization by the United States and other foreign producers, but its commitment to developing a WTO-compliant national semiconductor industry remains apparently robust.[24] Meanwhile, the Lenovo (formerly Legend) Group, another partially privatized, state-owned company (its largest originating shareholder is the Chinese Academy of Sciences) boasted $3 billion in revenue and a 30 percent share of China's PC market by 2003.[25] This domestic market, which in 2004 became the world's second-largest after the United States, was beset by intensifying competition and declining profit margins, so Lenovo was forced to look for growth farther afield. Already the world's eighth-largest computer maker, although overseas sales comprised merely 7 percent of revenues in 2002, the company formulated plans to push more aggressively into foreign markets.[26] In its home market as well as abroad, Lenovo faced a pack of well-established transnationals: Dell, Hewlett-Packard, IBM, Fujitsu, and Toshiba.[27] Although there is no certainty of success, Lenovo's next move was to acquire IBM's personal computer unit at the end of 2004. At a stroke, this granted Lenovo a worldwide presence and turned it into the world's third-largest computer maker.[28] Lenovo would continue to sell PCs under the IBM brand and was to relocate its world headquarters from Beijing to Armonk, New York, adjacent to IBM's own base. Lenovo's new CEO was also transferring from IBM—which, doubtless in hopes of leveraging the transaction to attain a more substantial presence in other Chinese information technology markets, acquired a minority ownership stake in Lenovo.[29] In the strategically important field of high-speed supercomputing, finally, Chinese progress again was striking. By 2004, China ranked with Germany as the world's fourth leading country in number of supercomputers.[30]

In Internet systems and services, efforts to build up leverage in the domestic market have once more combined with a new ability to raise capital via foreign stock exchanges to support increasing market power and transnational ambition. Huawei Technologies, a Chinese networking-equipment maker with twenty-four thousand employees and sales of $3.8 billion in 2004, is the most resounding success story. Huawei began "a significant U.S. push" in 2002, "touting its products' similarity to Cisco's market-dominant gear, at lower prices."[31] In 2003, Huawei teamed with 3Com to produce and sell networking gear to the business market;[32] and Huawei also announced its first contract, worth $900 million, to provide a commercial third-generation wireless network for multimedia services to Sunday Communications of Hong Kong.[33] Surviving a challenge from Cisco over intellectual property, by 2004 Huawei was selling its products in seventy countries, winning major contracts from Brazil to Sweden; it constitutes a serious competitor for beleaguered telecommunications manufacturers based in Europe and the United States.[34] Tom Online, a portal and purveyor of wireless content and online advertising in China, sought to raise $195 million on the Nasdaq Stock Market in the United States in March 2004; the top three Web portals at the time, Sohu.com, Sina.com, and Netease.com, were all profitable Nasdaq-listed companies.[35] A somewhat greater measure of international recognition has followed in the wake of China's Internet initiatives. ICANN, the U.S. organization that controls the naming of Web sites through the Internet's central registry, reconstituted itself in late 2002 in part to allow a Chinese representative onto its new advisory board. Whether or not China has found it feasible to "wield more influence over the Internet" as a result, as the *Financial Times* reported,[36] there is little question that it is trying to promote Internet-address standards that are less dominated by the United States.[37] China's fifty-nine million Internet users by the end of 2002 had already made it the world's fourth-largest Internet market, giving further assurance that its voice would be heard by policy-making bodies.[38]

In software and services, policies again are crafted to link transnational market expansion to domestic economic growth and continuing corporate reorganization. In March 1998, acting on general secretary Jiang Zemin's edict that "none of the four modernizations would be possible without informatization," China established a new "superministry" to promote information industries as a "new point of economic growth."[39] Individual Chinese software and services companies evince a protean character, as they seek to benefit from the process of "creative destruction" that the economist Joseph Schumpeter identifies with market forces. Chinadotcom, a continuing corporate experiment in information-market development, began as a Web service for the Xinhua press agency in 1997, then as an Internet portal partnership with AOL and a Nasdaq listing, and then sought to carve out a niche by purchasing distressed software companies in North America and Europe to develop a basis for selling software products to Chinese manufacturers.[40] China's software and services industry reportedly generated revenues of around $4 billion in 2003 (compared with $12 billion for its more visible Indian counterpart), with exports growing rapidly. "To spur the development of the software outsourcing industry," reported the *Financial Times,* "China has set up 15 national software industrial parks and is encouraging tertiary institutions to emphasize software development and applications in their curriculums."[41]

This leads us to yet another, increasingly important information market.

Private investment in China's schools and colleges commenced during the late 1980s and expanded alongside an explosion in demand for distinction-conferring educational services for middle- and upper-class children. In 2002, Chinese consumers were reported to have spent out-of-pocket $40 billion on education.[42] Further market growth is expected in the higher education segment, where, beginning in 1999, a "fee-increase mania" has set in. College administrators exploit every possible pretext to increase student fees, resulting in what one analyst calls "bottomless corruption, of a kind probably without precedent in the history of world education."[43] China's National People's Congress nevertheless ratified the shift by approving a law in December 2002 to grant private colleges parity with publicly supported institutions in possessing a right to make a "reasonable profit" on operations.[44] By 2005, Chinese universities were filing about six thousand patents annually—around the same total as U.S. universitites.[45]

These initiatives are too diverse and substantial to be mere happenstance. Do they possess an underlying political-economic coherence?

China's Rise: A Threat to U.S. Power?

Distinctions among different market segments are not only economic but also political, especially for the content industries and most particularly for urban press and broadcast news. In what Joseph Man Chan and Jack Linhuan Qiu call "a partially liberalized authoritarian media system," even as commercial and market pressures deepen their hold, the Chinese party-state continues to impose a formidable array of political and ideological controls.[46] The question we are engaging here, however, cuts in a different direction: do Chinese initiatives in communications and information possess any overarching pattern or directionality for the political economy of transnational capitalism?

Chinese officials are prone to cast these initiatives as a poor country's bootstrapping attempts at economic development; they claim to be merely redressing the glaring imbalances that continue to scar global society, including world communications. For example, the State Council information minister Zhao Qizheng has argued that "'Asian countries should set up their own strong media for the sake of speaking for themselves, reporting the facts about their countries, and speaking out to safeguard their national interests.'"[47]

Although China's leaders should not be praised for acting in principled support of democracy and equality, this perspective expresses an important truth: its far-ranging endeavors in this sector notwithstanding, China is nowhere near upending U.S. political-economic power, either in communications and information or in general. This does not signify, of course, that U.S. power will not decline for other reasons.

If Chinese leaders typically minimize their challenge, then U.S. policy makers are prone to overstate it by downplaying Chinese capital's mixed record of success to emphasize instead a looming threat to U.S. hegemony. "Thanks to dramatic progress in technology, transportation, and communications systems, China will wield far more power in the global economy," declared one of the milder statements of this kind in the influential journal *Foreign Affairs*.[48] What George J. Gilboy calls a new "mercantilist economic superpower" is said to be forming.[49]

U.S. state agencies, moreover, act consistently to neutralize or outflank what they deem to be incursions within the strategic information sector. It is significant that these state interventions are aimed not only at Chinese but also at U.S. capital. Important instances include prosecution by U.S. authorities of the U.S. aerospace companies Loral and Hughes Electronics for passing technical information about satellites to the Chinese;[50] locking out the successfully expanding $2 billion-per-year Chinese space program from the U.S.-dominated International Space Station;[51] rebuffing the Hong-Kong-based Hutchison from participating in the takeover of a global international telecommunications system possessing U.S. assets;[52] sending a strongly worded official letter protesting China's attempt to use technical standards to gain traction for its domestic companies in next-generation wireless-system markets;[53] and bringing a trade action against China for supposedly unfairly taxing semiconductor imports.[54]

Analysis must extend, however, beyond our list of initiatives and beyond a merely instrumental explanation of action and counteraction. Much of the significance of these events lies elsewhere, beyond any putative zero-sum game between current global hegemon and would-be rival. It is not merely that Chinese capital has achieved at best partial success in its early efforts to transnationalize, or that the U.S. authorities retain powerful forms of leverage over the global system. China's rise is profoundly intertwined with larger structural issues. The question becomes not simply how comprehensive or immediate a challenge one nation poses to the other, but what ongoing trends betoken for the structure and function of a more encompassing transnational capitalism. This vital point may be explicated by turning again to changes in telecommunications.

Telecommunications and China's Reintegration into Transnationalizing Capitalism

The Federal Communications Commission has documented the "steady growth in use of U.S. international facilities for international . . . private line services from the United States."[55] This bland formulation understates the significance of what has occurred. The pace of growth in international private-line services and the scale of the resulting reorganization of transnational business around them are unprecedented.

FCC data shows that, at the end of 2000, the number of activated circuits (measured in terms of sixty-four-kilobit-per-second equivalent circuits) was 2,178,926—a 121 percent increase from 1999[56]—and that it increased to 2,844,862 by the end of 2002.[57] The category of business-dominated leased circuits called International Private Line Services accounted for 70 to 74 percent of the total for each of the three years between 2000 and 2002, after having experienced disproportionately rapid growth over the preceding period at the expense of International Message Telephone Service—telephone calls transiting the traditional public switched network.[58]

In the case of individual country routes, the overall pace of network growth and the shift toward private lines are noteworthy. A longtime major U.S. trading partner, the United Kingdom, for example, became the United States's topmost telecommunications destination in 1999 (it was the third-largest in 1997) and has remained so; the number of activated circuits increased between 1997 and 2002 from 41,739 to 694,019, even as the proportion of voice-oriented public lines to private-line (inclusive of "other") circuits declined sharply, from .77 to .07.[59] For Canada, again a long-standing trade partner and the second top destination for U.S. callers, 95,481 activated circuits in 1997 became 391,449 in 2002, even as the ratio of public to proprietary lines declined from 1.12 to .32.[60] For Mexico, activated circuits increased from 60,555 to 233,261, while the public-private line ratio decreased from 1.57 to .36. And, exclusive of Hong Kong, activated circuits between the United States and China increased from 1,927 to 54,809, as China became an increasingly important country node (ranking between tenth and twelfth) for U.S. international telecommunications; the ratio of public to private circuits in this case declined especially precipitously, from 3.52 to .03.[61]

In China's case, this change was especially radical, as a brief retrospective look confirms. China detached itself from the U.S.-dominated international telecommunications system following the 1949 revolution. A small satellite station connecting the country to the Intelsat satellite network went into operation on a temporary basis during Richard Nixon's historic visit in 1972. Soon afterward, the Chinese signed multimillion-dollar contracts with three U.S. companies—RCA Global Communications, Communications Satellite Corporation, and Western Union International—to build a series of satellite ground stations, thereby kicking off the trend to increase the number of international links. At this juncture (early 1973), China was said to possess a mere eight telephone circuits to the United States, with telephone calls between the two nations running at about sixty-five a month. As recently as 1995, China and the United States were connected by only 1,114 activated circuits, of which private lines accounted for merely fifty-four—making the ratio of public to private circuits 19.59.[62] The spectacular buildup of international links to China—and, indeed, to much of the world—took place only as a consequence of the unprecedented telecommunications boom during the second half of the 1990s.

All told, for the top thirty international destinations that collectively claimed around 96 percent of total active circuits after the turn of the new century, the ratio declined from 1.03 in 1997 to .16 in 2002.[63] And these figures substantially *understated* the growth of proprietary lines, mainly because non–common carrier private cables were also newly authorized and built; their owners tried to sell the bulk of their capacity to end-users, notably Internet service providers and foreign carriers. These carriers were not required to report cable capacity to the FCC.[64]

Two trends combined to produce this sharply changed pattern. First, data applications carried by Internet service providers over leased or owned private lines grew markedly, at the expense of the traditional technology used for voice calling. Second, inhouse proprietary network applications proliferated among and between large organizations, principally transnational corporations.

While skyrocketing use of the Internet has contributed to this shift, it should be stressed that, during the overall network-building binge of the 1990s, perhaps two-thirds of total Internet investment was undertaken by businesses, to erect walled-off private systems.[65] These corporate-commercial intranets, as they were called, carry two types of information that together probably account for a substantial fraction of the hugely enlarged total of international private-line circuits: first, intra- and intercorporate data streams about production scheduling, inventories and supply chains, accounting, payroll, and research and development; second, an overlapping series of tradeable business services in finance, law, advertising, training, and other areas, which incorporate telecommunications as an intermediate input. These twin applications mostly did not fall within the domain of the open Internet—hence data concerning them were typically privileged—but rather within an explosively growing sphere of specialized, proprietary or quasi-proprietary network operations.

An additional point also needs to be made. In the Chinese context, the growth of international private-line circuits should not be seen as a purely instrumental projection, whether enacted mainly by foreign capital or by Chinese elites. Beyond the specific social agency that spearheads the buildout of any given private-network system lies the question of how this buildout structures or restructures the broader and more multifaceted market-expansion process.

Telecommunications networks function as a critical coordinating mechanism for dispersed corporate production and distribution chains. In this way, they also constitute a basic infrastructure for accelerated corporate transnationalization. Improved corporate access at lowered cost to international fiber-optic networks is a key historical prerequisite for reassimilating particular territories and reorganizing segments of the division of labor within a production system that is expanded by the same token: outsourced manufacturing and the offshoring of business and commercial services contribute to this process.[66]

The takeover of the regional network Asia Global Crossing by China Netcom must be seen not only as a direct bid for power by "China" but also as an attempt to participate profitably in the structural reorientation of the Chinese economy toward transnational capitalism. China Netcom displayed comparable ambitions by negotiating with PCCW to purchase a stake in the latter's HKT Telephone system in Hong Kong; the revealing strategic plan was to enlarge Netcom's market presence in adjoining Guangdong Province, where foreign-owned manufacturing plants are concentrated.[67] Struggles over the terms of China's accession into the World Trade Organization likewise must be interpreted on these same two planes: one that involves the play for advantage between particular industrial interests and states, and another that signifies programmatic expansion of linkages between China's domestic political economy and the transnational capitalism into which it is reintegrating.[68]

Processes of structural change are co-evolving with instrumental actions taken by a power bloc consisting, as Yuezhi Zhao specifies, "of the bureaucratic capitalists of a reformed Party state, transnational corporate capital, and an emerging urban middle class, whose members are the favored customers of both domestic and transnational capital"—and, for some important purposes, also of foreign state managers and policy makers.[69] On some of the most fundamental issues, crucially, these actors broadly agree, binding the Chinese national economy and its domestic social strata more and more comprehensively to the vicissitudes of a transnational market system.[70] Only the terms and conditions of this process of integration, and not the process itself, are objects of negotiation.

Viewed from the Chinese side, two interrelated dimensions of this structural metamorphosis stand out. First is the party state's somewhat undisciplined attempt to secure advantages for domestic capital, as its representatives broker transnational corporate access to the country's great reserves of cheap labor power and to its expanding domestic market. Second is its selective effort to build up individual domestic enterprises into effective transnational corporations.[71]

The first of these strategies is anything but novel. China is not the first country to try to leverage state control over the terms of entry into its domestic market in an effort to build up its economy and develop domestic capital. This strategy echoes a time-tested practice, pursued throughout prior decades with varying success by countries such as India and South Korea and earlier by Japan and the United States.[72] Inward foreign direct investment in China, however, is also being tied to a second strategic endeavor: to expand Chinese companies' outward foreign direct investment. In this respect, the contrast with earlier development plans is telling.

Economic-development initiatives of the mid-twentieth century limited inward and outward foreign direct investment and relied on import substitution (though sometimes more for rhetorical purposes than in reality[73]) and strategic linkages between emerging home industries on the grounds that *national* self-determination is best served by systematically reserving the domestic market for such guided initiatives. Whether state- or corporate-led, the endeavor was always framed in terms of national development. The $24 billion Three Gorges Dam project, and the nationalism that rhetorically infused it, might be cited as evidence of such an effort. But even this gigantic project must be set in the context of China's reliance—unparalleled for a poor country—on foreign direct investment, as Gilboy explains:

> Since it launched reforms in 1978, China has taken . . . ten times the total stock of FDI [foreign direct investment] Japan accumulated between 1945 and 2000. According to China's Ministry of Commerce, U.S. firms have invested more than $40 billion in more than 40,000 projects in China. Given its openness

to FDI, China cannot maintain its domestic market as a protected bastion for domestic firms, something both Japan and South Korea did during their periods of rapid growth. Instead, it has allowed U.S. and other foreign firms to develop new markets for their goods and services, especially high-value-added products such as aircraft, software, industrial design, advanced machinery, and components such as semiconductors and integrated circuits.[74]

Assimilating the structural logic of today's supranational market system requires China's policy makers to employ a conception of the national economy that holds hostage the social needs of the vast majority of the population to capital's demands for accumulation on a transnational scale. This tends to undercut any prospect of self-determination in the older, nation-bound sense and, nationalist rhetoric notwithstanding, to render this ideal less relevant.[75]

Because developing communications and information infrastructures have come to constitute a leading edge of transnational capital's overall accumulation project, this same strategic goal has likewise come to be internalized by Chinese leaders. The unprecedented buildout of a high-tech telecommunications grid—requiring larger capital investments than those funneled into the Three Gorges Dam—is a primary exemplar. Another was announced in 1997, when president Jiang Zemin stated, "Science and technology being a primary productive force, their progress is a decisive factor in economic development. We must . . . make the acceleration of their progress a vital task in economic and social development . . . strengthen basic research and research in high technology and accelerate the pace of applying high technology to production."[76] According to this developmental vision, incorporated into the Constitution of the Chinese Communist Party five years later at the Sixteenth Party Congress in November 2002, the party "must persist in taking economic development as the central task [and] give full play to the role of science and technology as the primary productive force."[77]

The obstacles in the way of building a high-tech informational basis for transnational capital—whether of Chinese or foreign origin—remain great. In one estimate, based on the number of original research papers published by biologists in internationally refereed "high-impact" journals, there were a mere five hundred productive biologists in China in 2003, compared with three thousand productive biologists of Chinese descent and forty thousand overall in the United States.[78] China's research and development budget for 2001 was $12.5 billion, when that of the United States was $281 billion.[79] And China continues to run a deficit on its technology trade with the world.[80]

But changes are apparent. Between 1981 and 2003, China increased its publications in international scientific journals twentyfold, accounting for over 5 percent of the world's scientific publications; in selected fields such as materials science, mathematics, and physics, its share was higher.[81] In the early 2000s, U.S. colleges graduated sixty thousand engineers each year; China's universities graduated two hundred thousand.[82] Again akin to India, China is making increasing efforts to entice foreign transnational corporations to send offshore to China back-office information services, as well as basic research, engineering, and design, and even financial analysis. Philips has shifted research and development on most televisions, mobile phones, and audio electronics to Shanghai, where General Electric performs important research and development work as well. Microsoft is spending $750 million over three years on research and development and outsourcing in China.[83] Intel, Motorola, and other high-tech transnational corporations set up over a hundred research and development centers, mostly in Shanghai and Beijing, to draw on the pool of technical and research talent and to sell more effectively into the Chinese domestic market.[84] Drug and biotech companies such as Roche and Pfizer have likewise found in China a quickly developing infrastructure, replete with "highly educated scientists who work for a fraction of what their Western counterparts are paid."[85] According to an official source, the world's largest transnational companies had established nearly four hundred research and development centers in China by 2001.[86] From the perspective of foreign capital, the lure of domestic demand and the availability of cheap scientific labor, linked by networks to transnational production chains, accounts for these moves.[87]

On the Chinese side, in contrast, the effort is to leverage control over inward foreign direct investment to build a globally competitive Chinese capitalism that can hold its own in the global information industry. In a progress report offered to the Davos World Economic Forum in January 2003, the president of China Netcom, Edward Tian, emphasized that the Chinese Communist Party Congress of November 2002 had adopted a "'very, very important policy—that in the next ten years China has to build . . . an information-led new economy.'"[88] Declaring that "'China can soon become the world's largest Internet and information economy,'" Tian asserted that, over the next decade, Chinese investment is likely to focus "'more to the software side and service sector,'" with the intention to "'export not only low-cost labor-intensive goods but . . . software and services to the Western world.'"[89] This formulation prompts additional questions about the systemic character of the structural transformation that is under way.

Regional or Global Integration?

Foreign transnational corporate contributions to the Chinese economic miracle have been manifold. China's export growth is remarkable not only for its scale—from $26 billion in volume in 1985 to $380 billion in 2003—but also for its strong linkages to foreign direct investment. Imports, particularly from Asian countries, are also strongly linked to foreign direct investment, as they are used for the reprocessing and export of finished goods.[90] From virtually none in 1984, foreign direct investment had cumulated to around $400 billion by 2001.[91] Over the next two years, China attracted an additional $100 billion or more in foreign direct investment, more than any other country, including the United States, and during 2004 and 2005 a further $120 billion flowed in.[92] By 2003, overall foreign direct investment accounted for no less than 40 percent of Chinese gross domestic product.[93]

Crucial questions must be posed, however, in regard to how this investment is aligning China within the system of transnational capitalism. Is China a general-purpose host for foreign direct investment from anywhere and everywhere? Is it moving into the orbit of one or another leading economic power? Or is it successfully transitioning into a transnational economic force in its own right?

The promise of market entry for foreign capital is tied to a buildup of select Chinese companies into transnational corporations. The goal is to try and transform a group of perhaps thirty to fifty state-owned enterprises, selling everything from oil and gas to white goods, steel and aluminum to PCs, TVs, and cell-phone service, into globally competitive units of capital by 2010.[94] There is evidence that this effort has already been partially successful. Eleven Chinese companies ranked in the Fortune 500 list of leading global enterprises by revenue.[95] A survey of around a hundred investment-promotion agencies by the United Nations Commission on Trade and Development in 2004 ranked China fifth in the world—ahead of sixth-place Japan—as a prospective source of investment capital.[96] Huawei is probably the outstanding example of a Chinese company that has rapidly established itself in overseas markets for advanced telecommunications and Internet technology.

Because a well-integrated national economy remains in doubt, owing in part to its problematic interconnections with transnational capital, this challenge should not be cast merely as a scramble to establish "national champions."[97] Thus far, at least, Chinese companies have notably "failed to develop strong domestic technology supply networks."[98] Nor should the extent of Chinese capital's global competitiveness be overstated. The difficulties of competing against long-established transnational behemoths are widespread and acute. Foreign-based transnational corporations themselves, furthermore, generate a large proportion of Chinese exports and domestic sales in key market segments. Between 1991 and 2001, the overall share of Chinese exports claimed by domestic affiliates of transnational corporations rose from 17 to 50 percent.[99] During the first half of 2003, by one estimate, no less than 84 percent of Chinese exports, by value, were by non-Chinese companies operating there.[100] In high-technology exports, whose share in China's overall trading volume increased from 3 to 22 percent between 1985 and 2000, foreign affiliates claim an especially dominant role. In electronic circuits, foreign affiliates took 91 percent of Chinese exports in 2000; in automatic data-processing machinery, they accounted for 85 percent of exports; and in mobile phones, foreign affiliates claimed 96 percent of China's exports in 2000.[101] Overall, 80 percent of Chinese technology imports and exports were controlled by foreign-owned firms.[102] "In contrast," emphasized a report by the United Nations Commission on Trade and Development, "Chinese domestic enterprises predominate in the low-technology sector, especially in the export of toys, travel bags, and yarns and fabrics."[103]

What about regional integration tendencies? China is certainly becoming robustly interconnected with Japanese capital and the newly industrialized East Asian economies. In 2000, Japan supplied 23.7 percent of China's imports of manufactured goods, while Hong Kong added 4.9 percent, and the rest of Asia (excluding West Asia) an additional 33.1 percent. The United States contributed only 12.2 percent, and the European Union 16.8 percent.[104] China ran trade deficits with Japan, Taiwan, Korea, and members of the Association of South East Asian Countries (ASEAN).[105] East Asian economies have correspondingly become more reliant on Chinese demand for their own growth, a fact that worries policy makers in, for example, the Republic of Korea.[106] Beijing has struck a free-trade pact with ten Southeast Asian nations and established an annual East Asian summit with the ASEAN countries, Japan, South Korea, India, New Zealand, Australia—but *not* the United States.[107] Some strategic moves in the information sector are positioned within this same regional matrix.

The Japan Information Technology Services Industry Association, the China Software Industry Association, and the Federation of Korean Information Industries agreed in 2003 to work together and with their respective governments to promote East Asia's use of open-source software in preference to continued reliance on Microsoft.[108] (This regional collaboration presumably arose in part from Japan's experience during the late

1980s, when it attempted to launch an alternative to Microsoft's proprietary operating system. Japan's effort was curtailed following strenuous U.S. government objections on the grounds that it would constitute a "trade barrier" to U.S. companies, that is, to Microsoft.[109]) In a related development, a Chinese computer scientist employed by a Hong Kong company, Culturecom Holdings, has developed a computer chip capable of responding directly to Chinese and other Asian languages, thereby potentially freeing the PC from microprocessors and operating systems programmed in English and the companies—Intel and Microsoft—that benefit most from this dependence.[110]

But a self-contained regional system under either Japanese or Chinese direction will be difficult to achieve. Centrifugal forces work against such a development. U.S. policy for East Asia aims to prevent cooperation between the two powers,[111] while Japanese leaders' unrepentant stance over Japanese imperial aggression throughout the first half of the twentieth century continues to inflame resentment throughout China.[112] Neither, however, do evolving economic relationships lend themselves to such a monochromatic regional vision. Hewlett-Packard, a leading PC manufacturer with a 17 percent share of the global market, responded to the East Asian open-source initiative in 2004 by announcing that it would sell PCs in Asia loaded with Linux by a Japanese software maker, Turbolinux.[113] Not only did U.S. and European transnational corporations continue to invest on an unprecedented scale in China, but Japanese, as well as Korean, Taiwanese, and other Asia-based foreign direct investment in China itself is based on the strategy of exporting to third markets, including, preeminently, the United States. The United Nations Commission on Trade and Development emphasizes that the aim of foreign direct investment in China by investors in East Asia is to "use China as an export platform for the Western markets."[114] On the one side, China—replacing the United States in this role—has become the biggest exporter to and, shortly afterward, the biggest trading partner with Japan.[115] On the other side, China overtook Japan in 2002 to become the third-largest exporter to the United States (after Canada and Mexico), sending $125 billion in exports.[116] China's national foreign exchange reserves have undergone a concurrent and dramatic increase to $258.6 billion by September 2002, the second-largest total in the world, then to $416 billion by early 2004, and $600 billion by early 2005.[117] China still ranks behind Japan (which increased its purchases of U.S. Treasury bills and bonds from $318 billion in December 2001 to $577 billion by January 2004, by which time it was financing about 40 percent of net U.S. Treasury-market borrowing), but Chinese investors have quickly become the second-largest force in U.S.-government debt markets—their purchases grew from $77 billion in December 2001 to $148 billion by January 2004.[118] In both cases, the attempt is to shore up the buying power of U.S. consumers, which sustains these Asian countries' own export growth.[119] There is at least some evidence of reciprocity: in the information and communications sector, U.S. technology companies such as Cisco, Google, Microsoft, Yahoo, Skype, and Sun Microsystems have helped equip Chinese authorities with hardware and software to restrict, filter, and monitor Chinese Internet users.[120]

Proliferating linkages between China and the world political economy transcend any bipolar configuration. As was also true for the United States, China's regional *and* global market-development initiatives are proceeding apace.[121]

Will China Provide an Escape Valve for the Global Crisis of Overproduction?

The two poles of market growth with which we are concerned share a common origin: capital's deep-seated need, in the face of growing worldwide overproduction throughout conventional industry, to identify and exploit new accumulation sites.[122] On the one side, the intensive cultivation of the emerging information sector of the economy arose in response to the 1970s-era profit slowdown and stagnation throughout the existing economy. Conditioned by this underlying pressure, on the other side, is the process of reintegrating China into what has become a more fully transnational capitalism. Because existing markets were glutted, the prospect of developing a vast new market in China for consumer goods and services constitutes an irresistible temptation. The competitive advantages to be derived from sourcing production in China's ultra-low-wage labor market has proven equally compelling, when rivals threaten to gain them first.

As Chinese companies and affiliates of transnational companies install production lines and infrastructure, and as distribution systems are built, mountains of commodities are sent out for sale into China's own domestic market and the greater world economy. Is China's national market capable of absorbing a greater share of this huge expansion of output? Or is China's integration likely to further destabilize transnational capitalism by worsening an already stressful condition of global overproduction?

Chinese policy makers have tried to nourish the soil in which capital grows by encouraging bank lending and easy credit; these constitute one of the major props of the country's massive capital spending and rapid economic growth.[123] In the consumer loan market, bank-card use has proliferated, and transnational market entrants like Citibank have tried to enlarge the market for consumer credit.[124] China's total debt outstanding in 2003 equaled nearly 160 percent of its economic output.[125] Room for maneuver via debt- and credit-fueled expansion therefore appears to be limited. But this does not exhaust the issue.

In March 2004, Premier Wen Jiabao conceded, "Deep-seated problems and imbalances in the economy over the years have not been fundamentally resolved."[126] An astonishing 90 percent of China's manufacturing product lines, the financial press estimated in 2004, were "in oversupply," "yet investment in fixed assets ... grew by 30% and contributed 47% of GDP" during 2003.[127] This continuing investment binge by domestic lenders and foreign companies threatens to become dangerously unhinged from actual sales and profits—by 2004, nearly 20 percent of total borrowings were officially accounted as nonperforming.[128] Existing overcapacity is all but certain to increase, sometimes sharply. During 2003, for example, General Motors, Volkswagen, Toyota, Honda, Nissan, Peugot Citroen, DaimlerChrysler, and Ford collectively announced more than $20 billion of *new* investment in China, enough to ensure that, for the foreseeable future, production of vehicles for the Chinese market will provide nearly twice as many autos as can be sold there.[129] As China becomes "a tough place for global auto makers," the auto makers are joining their fledgling Chinese rivals such as SAIC, Changan Automobile, and Dongfeng Motor (bulked up by a decade of growth via their partnerships with different transnational auto groups) in trying to export China-made vehicles to Europe and North America. This vital industry shift is therefore motivated as much by the companies' inability to expand the sated Chinese market as by corporate strength and robust ambition.[130]

At the same time, the rapid reintroduction of capitalist social relations aggravates uneven regional and social development and generates deepening inequality, profiteering, the breakdown of social-services and public-health infrastructure, and rising unemployment throughout this heavily agricultural nation.[131] The countryside is home to seven-tenths of China's 1.3 billion people but only half the nation's employment; in these rural districts, where farmers continue to be forced from their land, perhaps two hundred million people are "mobile"—a euphemism meaning that they migrate continually in search of work.[132] Joining this "floating population," additional tens of millions of work-seekers tramp the streets of China's cities.[133] In these circumstances, authorities are confronted by continual outbreaks of disorder and dissent.[134] In response, government policy has begun to notice rural China, and the leadership has recognized a tactical need to direct more state spending to the countryside. Once more, however, even if the will to do so is manifest—which remains far from clear—the room for maneuver is limited. An unusually comprehensive review of the condition of China's vast peasant class found that "over half the Chinese population falls into the World Bank's 'Fourth World' category, with the per capita purchasing power of the lowest-income countries."[135] No cohesively unified national market yet exists, however, and although concerted efforts to create one are evident, this is sure to take years of sustained effort. It would be foolish to forecast that limits placed on China's surplus absorption capabilities by rural poverty and a fragmented national market will be summarily swept aside.

Conclusion

Communications and information comprise a vital economic sector whose rapid expansion reflects capital's underlying need to arrest stagnation and profit decline. A second growth engine, born of the same compulsion, has materialized around China's reinsertion into the transnational market system. These two poles of growth are related, and their junction involves something quite different than the sum of their parts. China's takeoff into sustained growth has found a substantial but also complex expression in information and communications, while the tendency of manufacturers relocated to China to "dictate global prices of everything from steel to microchips"[136] seems likely to accentuate the continuing problem of overproduction in China and throughout the world economy at large. If so, then, paradoxically, the successful exploitation of these two poles of growth will contribute to a resurgence of the very economic crisis that prompted their own prior development.

Notes

Abbreviations Used in Notes

FT   *Financial Times*
NYT   *New York Times*
WSJ   *Wall Street Journal*

Preface

1. Nicholas G. Carr, *Does IT Matter?* (Boston: Harvard Business School Press, 2004).
2. Doug Henwood, *After the New Economy* (New York: New Press, 2003).

Chapter 1: How to Think about Information

1. American Express, advertisement, WSJ, 25 October 1982, 11.
2. Forrest Woody Horton Jr., "Rethinking the Role of Information," *Government Computer News* (April 1983): 1.
3. Joan Spero, "Information: The Policy Void," *Foreign Policy* 48 (Fall 1982): 152.
4. *Oxford English Dictionary*, vol. 5 (Oxford: Clarendon Press, 1933), 274.
5. *Oxford English Dictionary Supplement*, vol. 2 (Oxford: Clarendon Press, 1976), 300–301.
6. Norbert Wiener, *Cybernetics: Or, Control and Communication in the Animal and the Machine* (Cambridge: Massachusetts Institute of Technology Press, 1961), 11.
7. Claude Shannon and Warren Weaver, *The Mathematical Theory of Communication* (Urbana: University of Illinois Press, 1959); Warren Weaver, "The Mathematics of Communication," *Scientific American* 181.1 (1949): 11–15.
8. Klaus Krippendorff, "Information, Information Society, and Some Marxian

Propositions," Paper presented at the Thirty-Fourth Annual International Communications Association Conference, San Francisco, 24–28 May 1984.

9. Herman R. Branson, "Information Theory and the Structure of Proteins," in *Essays on the Use of Information Theory in Biology,* ed. Henry Quastler (Urbana: University of Illinois Press, 1953), 84.

10. Heinz von Foerster, Margaret Mead, and Hans Lukas Teuber, eds. *Cybernetics: Circular, Causal, and Feedback Mechanisms in Biological and Social Systems; Transactions of the 10th Conference, 22–24 April 1955, Princeton* (New York: Josiah Macy Jr. Foundation, 1956), 70.

11. Richard C. Raymond, "Communication, Entropy, and Life," *American Scientist* 38 (1950): 278.

12. James Watson and Francis Crick, "Genetical Implications of the Structure of Deoxyribonucleiic Acid," *Nature* 171 (1953): 965; Edward Yoxen, *The Gene Business* (New York: Harper and Row, 1984), 18.

13. Ten renowned conferences on "circular causal and feedback mechanisms in biological and social systems," or "cybernetics," sponsored by the Josiah Macy Jr. Foundation, did much to familiarize researchers from engineering, mathematics, physiology, psychiatry, anthropology, chemistry, philosophy, and other fields with emerging concepts of information theory. A considerable portion of the conferees' time was spent attempting to discern the relevance of notions of control, feedback, homeostasis, and other concepts associated with information theory for their host fields. This early conviction that the search for an informational component of analysis might be extended usefully to any system facilitated the development of increasingly precise knowledge of the informational workings of diverse systems. To biology, especially, information theory has proven a powerful stimulus to improved understanding. And growing knowledge of biological information processing suggests that such "natural systems" may offer vital clues to problems currently faced by designers of "artificial" information networks. As one researcher puts it, "processing is ubiquitous in natural systems," and computer scientists seek to deepen understanding of "architecture, data structures, and algorithms" by "viewing them in both natural and artificial settings." J. R. Sampson, *Biological Information Processing: Current Theory and Computer Simulation* (New York: John Wiley and Sons, 1984), 4, 5.

14. Stafford Beer, "The Irrelevance of Automation," *Cybernetica* 1.4 (1958): 286. "The most complex situations we have heard discussed are the stabilities engendered by inverse feedback in social structures of isolated communites reported principally by social anthropologists" (von Foerster, Mead, and Teuber, eds., *Cybernetics,* 74).

15. Anthony Oettinger, "Information Resources: Knowledge and Power in the 21st Century," *Science* 209 (4 July 1980): 192.

16. Krishan Kumar, *Prophecy and Progress: The Sociology of Industrial and Post-Industrial Society* (New York: Penguin, 1978).

17. Harlan Cleveland, *The Knowledge Executive* (New York: Dutton, 1985), 20.

18. Ibid., 33.

19. Ibid., 25, 34, 29.

20. A fourth tradition of thinking about information did in fact emerge as an outgrowth of conventional economic theory, as more businesses began to confront issues associated with markets for network systems and information services. The idea of information as a public good may be placed within this framework. Economists including Kenneth Arrow, Kenneth Boulding, Fritz Machlup, Marc Porat, and Joseph Stiglitz elaborated this conception, which overlapped with the postindustrialists' treatment, providing added support for the idea that an emerging information-intensive economy would incarnate "new rules" or at least require specific adaptations of existing economic doctrine. For a recent discussion, see Council of Economic Advisors, "The Role of Intellectual Property in the Economy," chap. 10 of *Economic Report of the President,* transmitted to the Congress February 2006, 211–30; retrieved 22 February 2006, www.whitehouse.gov/cea/pubs.html. For an influential revision and synthesis, see Carl Shapiro and Hal R. Varian, *Information Rules: A Strategic Guide to the Network Economy* (Boston: Harvard Business School Press, 1999).

21. Jane Yurow, *Issues in Information Policy,* Department of Commerce Publication No. NTIA-SP-80–9 (Washington, D.C.: Government Printing Office, 1981), 54.

22. Rudolf Hilferding, "Bohm-Bawerk's Criticism of Marx," in *Karl Marx and the Close of His System: Bohm-Bawerk's Criticism of Marx,* ed. Paul Sweezy (Philadelphia: Orion, 1984), 132–33.

23. Paul Baran and Paul Sweezy, *Monopoly Capital: An Essay on the American Economic and Social Order* (New York: Monthly Review, 1966), 125, 122.

24. Sean Wilentz, *Chants Democratic: The American Working Class, 1788–1850* (New York: Oxford University Press, 1984).

25. Baran and Sweezy, *Monopoly Capital,* 127.

26. Karl Marx, "Results of the Immediate Process of Production," in *Capital,* vol. 1., trans. Ben Fowkes (New York: Vintage, 1977), 1044.

27. Eric Roll, *A History of Economic Thought,* 3d ed. (Englewood Cliffs, N.J.: Prentice-Hall, 1956), 129.

28. Marx, "Results of the Immediate Process of Production," 1048.

29. Nicholas Garnham, "Contribution to a Political Economy of Mass-Communication," *Media, Culture, and Society* 1 (1979): 123–46.

30. Marx, "Results of the Immediate Process of Production," 1044. See also David Harvey, *The Limits to Capital* (Chicago: University of Chicago Press, 1982), 104–5.

31. Paul Sweezy and Harry Magdoff, "The Strange Recovery of 1983–1984," *Monthly Review* 37.5 (1985): 1–11.

32. U.S. Department of Commerce, *Survey of Current Business* 66.7 (Washington, D.C.: Government Printing Office, 1986), 60, table 5.7; U.S. Department of Commerce, *The National Income and Product Accounts of the United States, 1929–82* (Washington, D.C.: Government Printing Office, 1986), 234–35, table 5.7.

33. Ralph Winter, "Forecast for '86 Capital Outlays Improves," WSJ, 30 April 1986, 6.

34. Sweezy and Magdoff, "Strange Recovery of 1983–1984," 6.

35. Ibid., 6, 7.

36. Ibid., 9.

37. Ibid., 9.

38. Wayne J. Howe, "The Business Services Industry Sets Pace in Employment Growth," *Monthly Labor Review* 109 (1986): 29, 33.

39. Ibid., 34. It should be noted that the business services category lumps together disparate kinds of jobs, including computer experts and janitors, management consultants and security guards. Nevertheless, the segment of the business services category devoted to information-intensive services—and it is a substantial component, probably more than half the total—registered employment gains generally far in excess of those characteristic of the overall economy.

40. Ernest Mandel, *Late Capitalism* (London: Verso, 1978).

41. Dallas Smythe, "Communication: Blindspot of Western Marxism," *Canadian Journal of Political and Social Theory* 1.3 (1977): 1–27.

42. "Economic Outlook," Testimony of Chairman Alan Greenspan before the Joint Economic Committee, U.S. Congress, 3 November 2005, 3; retrieved 13 January 2006, www.federalreserve.gov/boarddocs/testimony/2005/20051103/default.htm.

Chapter 2: Culture, Information, and Commodification

1. Steven J. Heims, *The Cybernetics Group* (Cambridge: Massachusetts Institute of Technology Press, 1991).

2. Abbe Mowshowitz, "On the Market Value of Information Commodities: I. The Nature of Information and Information Commodities," Management Report Series no. 90 (Rotterdam, Netherlands: Erasmus University, Rotterdam School of Management, 1991), 5.

3. Wilbur Schramm, "Information Theory and Mass Communication," *Journalism Quarterly* 32.5 (1955): 135.

4. David A. Mindell, *Between Human and Machine* (Baltimore: Johns Hopkins University Press, 2002), 136–37.

5. Quoted in Arthur L. Norberg, *Computers and Commerce: A Study of Technology and Management at Eckert-Mauchly Computer Company, Engineering Research Associates, and Remington Rand, 1946–1957* (Cambridge: Massachusetts Institute of Technology Press, 2005), 185.

6. L. David Ritchie, *Information* (Newbury Park, Calif.: Sage Publications, 1991), 7, 31.

7. Ibid., 47.

8. For this intellectual sea change, see Dan Schiller, *Theorizing Communication: A History* (New York: Oxford University Press, 1996).

9. For a spectacular example, purporting to present an integrated analysis of the roles of information, matter, and energy, in "systems" ranging from cell to society, see James G. Miller, *Living Systems* (New York: McGraw-Hill, 1978). For a similar example of shifting without apparent effort or concern across these levels, see James Beniger, *The Control Revolution* (Cambridge, Mass.: Harvard University Press, 1986).

10. Mowshowitz, "On the Market Value of Information Commodities," 6.

11. Klaus Krippendorff, "Paradox and Information," in *Progress in Communication Sciences*, vol. 5, ed. Brenda Dervin and Melvin J. Voigt (Norwood, N.J.: Ablex, 1984), 50.

12. Hans Christian von Baeyer, *Information: The New Language of Science* (Cambridge, Mass.: Harvard University Press, 2004), 25.

13. Daniel Bell, "The Social Framework of the Information Society," in *The Computer Age: A Twenty-Year View,* ed. Michael L. Dertouzos and James Moses (Cambridge: Massachusetts Institute of Technology Press, 1979), 168.

14. An African American packinghouse worker and trade unionist, interviewed at his home in Chicago in the mid-1960s, "walks over to the piano, removes the plastic cover, and noodles some roughhewn blues chords as he talks. 'I call this culture. That's my best definition of culture. When people are oppressed, sometimes they have to have some way. . . . [Mahalia Jackson] is a typical example of what I'm trying to say. Like when my mother died, her music made me cry, but it gave me hope.'" Studs Terkel, *Division Street America* (New York: New Press, 1993), 134–35.

15. Krishan Kumar, *Prophecy and Progress: The Sociology of Industrial and Post-Industrial Society* (Harmondsworth: Penguin, 1978).

16. Daniel Bell, "The Eclipse of Distance," *Encounter* 20.5 (1963): 564; Daniel Bell, *The Cultural Contradictions of Capitalism* (New York: Basic Books, 1976), xi, 12; Fred Block, *Postindustrial Possibilities: A Critique of Economic Discourse* (Berkeley: University of California Press, 1990), 7.

17. Dan Schiller, "The Legacy of Robert A. Brady: Antifascist Origins of the Political Economy of Communications," *Journal of Media Economics* 12.2 (1999): 89–102.

18. This usage is common in journalistic references, like this one, to the purported aversion of Japanese corporations to buying software systems off-the-shelf: "They want a system that's unique to them, not a commodity." Michael Schrage, "Software Powerhouses Remain Elusive Goal for Japanese," *Los Angeles Times,* 13 May 1993, 1D. For the president of RCA, Robert Sarnoff, the emergent economic role of information was already prominent by 1967, when he predicted that "'information will become a basic commodity equivalent to energy in the world economy,'" able to "'function as a form of currency in world trade, convertible into goods and services everywhere.'" Quoted in Herbert I. Schiller, *Mass Communications and American Empire* (New York: Augustus M. Kelley, 1969), 51.

19. Vincent Mosco, *The Political Economy of Communication: Rethinking and Renewal* (London: Sage, 1996).

20. Nicholas Garnham, *Capitalism and Communication* (London: Sage, 1990); Bernard Miege, *The Capitalization of Cultural Production* (New York: IMG, 1990); Mosco, *Political Economy of Communication*.

21. In territorial terms, it might be added, the unit of analysis has been usually implicit—the nation state. Moreover, the site of "the information economy" has been typically identified with the United States or, in some formulations, with the "developed market economies": the United States, Western Europe, and Japan. Where is the information commodity developing most intensively today? A national unit of analysis is inadequate to answer this question.

22. John Clarke, *New Times and Old Enemies* (London: HarperCollins Academic, 1991), 98; Garnham, *Capitalism and Communication,* 37.

23. Bob Rowthorn and Donald J. Harris, "The Organic Composition of Capital and Capitalist Development," in *Rethinking Marxism,* ed. Stephen Resnick and Richard Wolf (New York: Autonomedia, 1985), 345–57.

24. For an especially insightful study, see James Curran, "The Press as an Agency of Social Control," in *Newspaper History: From the Seventeenth Century to the Present Day,* ed. George Boyce, James Curran, and Pauline Wingate (Beverly Hills, Calif.: Sage Publications, 1978), 51–75.

25. Margaret Graham, *The Business of Research: RCA and the Videodisc* (New York: Cambridge University Press, 1986).

26. See Dan Schiller, *Objectivity and the News: The Public and the Rise of Commercial Journalism* (Philadelphia: University of Pennsylvania Press, 1981); Michael Denning, *Mechanic Accents* (London: Verso, 1986); Elliott J. Gorn, *The Manly Art: Bare-Knuckle Prize Fighting in America* (Ithaca, N.Y.: Cornell University Press, 1986); Kathy Peiss, *Cheap Amusements: Working Women and Leisure in Turn-of-the-Century New York* (Philadelphia: Temple University Press, 1986); Elizabeth Ewen, "City Lights: Immigrant Women and the Rise of the Movies," *Signs* 5.3 (1980; Supplement): S45–65; Roy Rosenzweig, *Eight Hours for What We Will* (New York: Cambridge University Press, 1983), 191–221; Steven J. Ross, "Struggles for the Screen: Workers, Radicals, and the Political Uses of Silent Film," *American Historical Review* 96.2 (1991): 333–67; Steven J. Ross, *Working-Class Hollywood* (Princeton, N.J.: Princeton University Press, 1997); George Lipsitz, *A Rainbow at Midnight: Class and Culture in Cold War America* (New York: Bergin and Garvey, 1982), 195–225; Michele Martin, "Capitalizing on the 'Feminine' Voice," *Canadian Journal of Communication* 14.3 (Spring 1989): 42–62; Michele Martin, *"Hello, Central?" Gender, Technology, and Culture in the Formation of Telephone Systems* (Montreal: McGill-Queen's University Press, 1991); William Boddy, "The Rhetoric and Economic Roots of the American Broadcasting Industry," *Cine-Tracts* 2.2 (Spring 1979): 37–54; Lynn Spigel, *Make Room for TV* (Chicago: University of Chicago Press, 1992); Lizabeth Cohen, *Making a New Deal: Industrial Workers in Chicago, 1919–1939* (New York: Cambridge University Press, 1990); Lizabeth Cohen, "The Class Experience of Mass Consumption," in *The Power of Culture: Critical Essays in American History,* ed. Richard W. Fox and T. J. Jackson Lears (Chicago: University

of Chicago Press, 1993), 135–60; Susan G. Davis, *Parades and Power: Street Theater in Nineteenth-Century Philadelphia* (Philadelphia: Temple University Press, 1986); Alexander Saxton, *The Rise and Fall of the White Republic: Class Politics and Mass Culture in Nineteenth-Century America* (London: Verso, 1990).

27. Herbert I. Schiller, *Culture, Inc.* (New York: Oxford University Press, 1989), 32. See also Vincent Mosco, *The Pay-Per Society: Computers and Communication in the Information Age* (Toronto: Garamond, 1989); Mosco, *Political Economy of Communication.*

28. Harry Braverman, *Labor and Monopoly Capital* (New York: Monthly Review, 1974).

29. Jim Davis and Michael Stack, "Knowledge in Production," *Race and Class* 34.3 (1993): 1–14.

30. JoAnne Yates, *Control through Communication* (Baltimore: Johns Hopkins University Press, 1989).

31. Braverman, *Labor and Monopoly Capital;* Harley Shaiken, *Work Transformed: Automation and Labor in the Computer Age* (New York: Holt, Rinehart, and Winston, 1983); S. Cohen, "The Labor Process Debate," *New Left Review* 165 (September–October 1987): 34–50.

32. Tessa Morris-Suzuki, "Robots and Capitalism," *New Left Review* 160 (November–December 1984): 120–21; David Dickson, *The New Politics of Science* (New York: Random House, 1983).

33. Garnham, *Capitalism and Communication,* 20–55.

34. Edward Yoxen, *The Gene Business* (New York: Harper and Row, 1983), 198.

35. Ibid., 19, 18.

36. Martin Kenney, *Biotechnology: The University-Industrial Complex* (New Haven, Conn.: Yale University Press, 1986), 4.

37. Jack R. Kloppenberg Jr., *First the Seed: The Political Economy of Plant Biotechnology, 1492–2000* (Cambridge: Cambridge University Press, 1988); Manuel Castells, *The Informational City* (Oxford: Blackwell, 1989), 12; Bronwyn Parry, *Trading the Genome: Investigating the Commodification of Bio-Information* (New York: Columbia University Press, 2004), 9.

38. Dan Schiller, *Digital Capitalism: Networking the Global Market System* (Cambridge: Massachusetts Institute of Technology Press, 1999).

39. Raymond Williams, *Keywords,* rev. ed. (New York: Oxford University Press, 1983), 87.

40. "Genetically Modified Food Items Are Common, but Little Noticed," WSJ, 24 March 2005, D4.

41. Kloppenburg, *First the Seed,* 152–90. See also Jack R. Kloppenburg Jr. and David L. Kleinman, "Seed Wars," *Socialist Review* 95 (1987): 7–41.

42. Walter Benjamin, "The Work of Art in the Age of Mechanical Reproduction," in *Illuminations,* ed. Hannah Arendt, trans. Harry Zohn (New York: Shocken, 1969), 217–51.

43. Daniel Bell, "The Third Technological Revolution," *Dissent* 36.2 (Spring 1989): 164–76. Yates (*Control through Communication*) makes a more illuminating, if still conceptually problematic, effort to give postindustrial society a reputable past by marrying Alfred D. Chandler's managerially focused business history to the corporate applications of earlier generations of information technology between 1840 and 1920. Beniger (*Control Revolution*) appropriates a thinly disguised Weberian notion of rationalization to argue for a long historical sequence of revolutionary changes beginning, incongruously, with a minor industrial accident in the 1840s.

44. Michel Foucault, "What Is an Author?" in *Textual Strategies*, ed. J. V. Harari (Ithaca, N.Y.: Cornell University Press, 1979), 141–60; Martha Woodmansee, "The Genius and the Copyright: Economic and Legal Conditions of the Emergence of the 'Author,'" *Eighteenth-Century Studies* 17.4 (1984): 425–48; Peter Jaszi, "Toward a Theory of Copyright: The Metamorphosis of 'Authorship,'" *Duke Law Journal* 40 (1991): 455–502; "Intellectual Property and the Construction of Authorship," *Cardozo Arts and Entertainment Law Journal* 10.2 (1992): 277–725.

45. U.S. Congress, Office of Technology Assessment, *Intellectual Property Rights in an Age of Electronics and Information*, OTA-CIT-302 (Washington, D.C.: Government Printing Office, 1986); U.S. Congress, Office of Technology Assessment, *Finding a Balance: Computer Software, Intellectual Property, and the Challenge of Technological Change*, OTA-TCT-527 (Washington, D.C.: Government Printing Office, 1992); Ronald V. Bettig, *Copyrighting Culture: The Political Economy of Intellectual Property* (Boulder, Colo.: Westview, 1996); Chakravarthi Raghavan, *Recolonization: GATT, the Uruguay Round, and the Third World* (London: Zed, 1990); Siva Vaidyanathan, *Copyrights and Copywrongs: The Rise of Intellectual Property and How It Threatens Creativity* (New York: New York University Press, 2001).

46. Jane Gaines, *Contested Culture* (Durham, N.C.: Duke University Press, 1991), 67, 65.

47. Martha Woodmansee, "The Genius and the Copyright: Economic and Legal Conditions of the Emergence of the 'Author,'" *Eighteenth-Century Studies* 17.4 (1984): 425–48.

48. James Boyle, "A Theory of Law and Information: Copyright, Spleens, Blackmail, and Insider Trading," *California Law Review* 80.6 (December 1992): 1469.

49. Mark Rose, *Authors and Owners: The Invention of Copyright* (Cambridge, Mass.: Harvard University Press, 1993).

50. Boyle, "Theory of Law and Information," 1469.

51. Jeanne T. Allen, "Copyright and Early Theater, Vaudeville, and Film Competition," *Journal of the University Film Association* 31.2 (1979): 6–11.

52. Lyman Ray Patterson, *Copyright in Historical Perspective* (Nashville: Vanderbilt University Press, 1968).

53. Rose, *Authors and Owners.*

54. For a study that insightfully extends this tradition of analysis, see Carla Hesse,

Publishing and Cultural Politics in Revolutionary Paris, 1789–1810 (Berkeley: University of California Press, 1991).

55. Harry I. Dutton, *The Patent System and Inventive Activity during the Industrial Revolution, 1750–1852* (Manchester: Manchester University Press, 1984); Doron S. Ben-Attar, *Trade Secrets: Intellectual Piracy and the Origins of American Industrial Power* (New Haven, Conn.: Yale University Press, 2004).

56. Joel Mokyr, *The Gifts of Athena: Historical Origins of the Knowledge Economy* (Princeton, N.J.: Princeton University Press, 2004), and especially Christopher May and Susan K. Sell, *Intellectual Property Rights: A Critical History* (Boulder, Colo.: Lynne Rienner, 2006).

57. An equivalent criticism may be directed against the claims of those who seek to ground parallel arguments on technology, freed, as in such cases it characteristically is, from the complex of productive relations that necessarily encases it, thereby allowing it to take real historical form. For an example discussed further in chapter 6, see Ithiel de Sola Pool, *Technologies of Freedom* (Cambridge, Mass.: Harvard University Press, 1983).

58. John Feather, *A History of British Publishing* (London: Routledge, 1988).

59. Allen, "Copyright and Early Theater," 11.

60. Ibid., 7.

61. Max Horkheimer and Theodor W. Adorno, *Dialectic of Enlightenment*, trans. John Cumming (New York: Seabury, 1972), 120–67; Ian Watt, *The Rise of the Novel* (1957; reprint, Berkeley: University of California Press, 1974).

62. Leo Lowenthal, *Literature, Popular Culture, and Society* (Palo Alto, Calif.: Pacific Books, 1968), 14–108.

63. Raymond Williams, *The Long Revolution*, rev. ed. (Harmondsworth: Pelican, 1965). A tension exists between Williams's avowed effort in the preface to interconnect changes in culture with these entrenched arguments for revolutionary shifts in economics and politics and his frequent (and effective) tendency to range freely back even to medieval times when framing specific assertions.

64. Stuart Hall, "Culture, the Media and the 'Ideological Effect,'" in *Mass Communication and Society*, ed. James Curran, Michael Gurevitch, and James Woollacott (London: Edward Arnold, 1977), 339.

65. Williams, *Long Revolution*, 290.

66. The historian Eric Hobsbawm helped to popularize the idea of the dual revolution. Hobsbawm, however, significantly insisted that "so profound a transformation cannot be understood without going back very much further in history than 1789." Yet, he concluded pragmatically, "How far back into history the analyst should go—whether to the mid-seventeenth-century English Revolution, to the Reformation and the beginning of European military world conquest and colonial exploitation in the early sixteenth century, or even earlier—is for our purposes irrelevant." Eric J. Hobsbawm, *The Age of Revolution, 1789–1848* (New York: Mentor, 1962), 18.

67. Neil McKendrick, John Brewer, and J. H. Plumb, *The Birth of a Consumer Society: The Commercialization of Eighteenth-Century England* (Bloomington: Indiana University Press, 1982).

68. Ellen Meiksins Wood, *The Pristine Culture of Capitalism* (London: Verso, 1991); Maurice Dobb, *Studies in the Development of Capitalism* (New York: International Publishers, 1963); T. S. Aston and C. H. E. Philpin, eds., *The Brenner Debate: Agrarian Class Structure and Economic Development in Pre-Industrial Europe* (Cambridge: Cambridge University Press, 1987); Rodney Hilton, *The Transition from Feudalism to Capitalism* (London: Verso, 1978); Peter Linebaugh, *The London Hanged: Crime and Civil Society in the Eighteenth Century* (London: Allen Lane, 1992); Edward P. Thompson, "The Peculiarities of the English," in *The Poverty of Theory and Other Essays* (London: Merlin, 1978), 35–91; Perry Anderson, "Origins of the Present Crisis," in *English Questions* (London: Verso, 1992), 15–47.

69. E. P. Thompson, *Whigs and Hunters* (New York: Pantheon, 1975).

70. Wood, *Pristine Culture of Capitalism*.

71. Or into postindustrial theory—which is perhaps predictable, given the latter theorists' sometime acquaintanceship with, and dogmatic aversion to, an earlier generation of Marxist historical scholarship. Bell attributes to Marx the idea that the working class equates exclusively with the industrial proletariat and then crows that the transition to a postindustrial society, with its large service-sector workforce, invalidates yet another aspect of the Marxian schema. If agricultural rather than industrial labor constituted the first capitalist workforce, then Bell's characterization loses any significance. Bell, "Third Technological Revolution," 168.

72. In Williams, we repeatedly meet versions of this progression, though rendered with unusual care. "The newspaper was the creation of the commercial middle class, mainly in the eighteenth century. . . . The fact is that the economic organization of the press in Britain has been predominantly in terms of the commercial middle class which the newspapers first served. When papers organized in this way reached out to a wider public, they brought in the new readers on a market basis and not by means of participation or genuine community relations." Williams, *Long Revolution,* 197, 210 (for some equally suggestive comments on the drama, see 292). Habermas's quite different concern for the emergence of "the public sphere" utilizes the same generation of scholarly sources in seeking to lodge its origins in reputedly decisive changes occurring at a range of social locations—the bourgeois press, the bourgeois family, and bourgeois state-society relations—again in eighteenth-century England. The public sphere is explicitly designated a category of bourgeois—rather than capitalist—society. Jurgen Habermas, *The Structural Transformation of the Public Sphere* (Cambridge: Massachusetts Institute of Technology Press, 1989).

73. Williams, *Long Revolution,* 271–99.

74. Lora E. Taub, "Enterprising Drama: The Rise of Commercial Theater in Early Modern London" (Ph.D. dissertation, University of California at San Diego, 1998).

75. Lucien Febvre and Henri-Jean Martin, *The Coming of the Book: The Impact of Printing, 1450–1800* (London: NLB, 1976); Feather, *History of British Publishing;* Margaret Spufford, *Small Books and Pleasant Histories* (Cambridge: Cambridge University Press, 1981).

76. Taub, "Enterprising Drama."

77. Williams, *Long Revolution,* 386.

78. Michael McKeon, *The Origins of the English Novel, 1600–1740* (Baltimore: Johns Hopkins University Press, 1987); Jeremy D. Popkin, *News and Politics in the Age of Revolution* (Ithaca, N.Y.: Cornell University Press, 1990).

79. Hall himself asserts, though for other reasons, that the search for a fixed inventory of cultural forms like the novel may be misguided. Stuart Hall, "Notes on Deconstructing 'The Popular,'" in *People's History and Socialist Theory,* ed. Raphael Samuel (London: Routledge and Kegan Paul, 1981), 235.

Chapter 3: Accelerated Commodification

1. "The stagnation in the world economy in the last few years has changed rapidly and dramatically the context of the discussion. . . . The economic role of the information and media industries and the services they provide are now primary factors in the maintenance of the material system of power, domestically and internationally." Herbert I. Schiller, *Who Knows: Information in the Age of the Fortune 500* (Norwood, N.J.: Ablex, 1981), xv. "These developments are in their turn determined by underlying economic trends and the efforts of transnationals, often backed by the economic planning instruments of nation states such as Japan and France, to develop the market for so-called information goods and services as a new growth sector. Since the late '60s electronics has been one of the key sectors of the world economy. . . . As productivity leveled off and profits dropped in more traditional manufacturing sectors and product lines the transnationals moved into the electronics sector in search of new products, new markets, and renewed growth." Nicholas Garnham, "Public Service versus the Market," *Screen* 24.1 (January–February 1983): 10. "It is . . . largely as a response to systemic crisis in the world business system that the new information technologies are being introduced into most of the developed market economies." Herbert I. Schiller, *Information and the Crisis Economy* (Norwood, N.J.: Ablex, 1984), 2. "[T]he information society concept . . . was an idea, devised by bureaucrats, academics, businessmen, and politicians, as a solution to a specific crisis of capitalism." Tessa Morris-Suzuki, *Beyond Computopia: Information, Automation, and Democracy in Japan* (London: Kegan Paul International, 1988), 58.

2. Robert Brenner, *The Boom and the Bubble: The U.S. in the World Economy* (London: Verso, 2002).

3. Peter Dicken, *Global Shift: Transforming the World Economy* (New York: Guilford Press, 1998), 28.

4. The Conference Board, *Information Technology Initiatives for Today: Decisions*

That Cannot Wait (New York: The Conference Board, 1972), 1. See Vincent Mosco, *Pushbutton Fantasies: Critical Perspectives on Videotex and Information Technology* (Norwood, N.J.: Ablex, 1982), 35–36.

5. U.S. Congress, House of Representatives, 96th Cong, 2d Sess., House Report No. 96–1535, "International Information Flow: Forging a New Framework," 32d Report by the Committee on Government Operations, 11 December 1980 (Washington, D.C.: Government Printing Office, 1980), 42. See Herbert I. Schiller, *Culture Inc.: The Corporate Takeover of Public Expression* (New York: Oxford, 1989), 118–19; Schiller, *Information and the Crisis Economy*, 2; Morris-Suzuki, *Beyond Computopia;* Simon Nora and Alain Minc, *The Computerization of Society: A Report to the President of France* (Cambridge: Massachusetts Institute of Technology Press, 1980).

6. Quotes from the statement of principles as given by Homer A. Jack, "Bandung: An On-the-Spot Description of the Asian-African Conference, Bandung, Indonesia, April 1955" (N.p.: Toward Freedom, n.d.), 2.

7. Dan Schiller, *Theorizing Communication: A History* (New York: Oxford, 1996), 102; Kaarle Nordenstreng, *The Mass Media Declaration of UNESCO* (Norwood, N.J.: Ablex, 1984), 69.

8. Kaarle Nordenstreng, "The MacBride Report: A Milestone in the Great Media Debate," Submission to Quadernos del CAC, Barcelona, Monograph on the 25th Anniversary of the Approval of the MacBride Report, April 2005: 1–2 (manuscript in author's possession).

9. Quoted in Nordenstreng, *Mass Media Declaration of UNESCO,* 74.

10. Quoted in ibid.

11. Vincent Mosco, *The Political Economy of Communication: Rethinking and Renewal* (London: Sage, 1996), 152; James W. Cortada, *The Digital Hand* (New York: Oxford University Press, 2004).

12. In France, the high point of concern developed in the interval between the publication of Jean-Jacques Servan-Schreiber's *The American Challenge* (London: Hamilton, 1968) and Nora and Minc's *Computerization of Society* (1978).

13. U.S. superiority persisted, but on a more qualified basis. In 2000, the United States accounted for half of all rich-country research and development expenditures in ICT manufacturing; Japan, the runner-up, claimed 21 percent. U.S. National Science Board, *Science and Engineering Indicators 2004,* vol. 1 (Arlington, Va.: National Science Foundation, NSB, 04–1), 4–60 (fig. 4–31); Antonio Regalado, "R&D Outlays to Rise in 2005, Driven by Military," WSJ, 7 January 2005, A2, A6. One government report boasted that "American-owned IT companies lead their foreign rivals in almost every segment of business activity—from research and development to design, production, and marketing." U.S. Department of Commerce, Economics and Statistics Administration, "Digital Economy 2003," 35; retrieved February 2006, www.esa.doc.gov/2003cfm. Consumer electronics, which Japanese upstarts including Sony and Matsushita had compelled an earlier generation of U.S. high-tech companies such as RCA and Magnavox to abandon, appear to have been reopened on a selective basis

to U.S. capital. Gary McWilliams, "In Electronics, U.S. Companies Seize Momentum from Japan," WSJ, 10 March 2005, A1, A10. Despite these claims, the U.S. comparative advantage in information—and the global dominance to which it contributed—was on the wane.

14. For a mainstream account that explicates some of the key themes, see Gerald Brock, *The Second Information Revolution* (Cambridge, Mass.: Harvard University Press, 2003).

15. Kenneth Flamm, *Creating the Computer* (Washington, D.C.: Brookings Institution, 1988); Janet Abbate, *Inventing the Internet* (Cambridge: Massachusetts Institute of Technology Press, 2000); Michael Lesk, *Understanding Digital Libraries,* 2d ed. (Amsterdam: Elsevier, 2005), 322–23.

16. Dan Schiller, *Digital Capitalism,* 1–88.

17. David Harvey continues to situate this tendency chiefly in light of its contributions to "space-time compression." David Harvey, *The New Imperialism* (Oxford: Oxford University Press, 2003), 98. For an approach that stresses the role of ICTs in augmenting "control," see Peter McMahon, *Global Control: Information Technology and Globalization since 1845* (London: Edward Elgar, 2002). Productivity statistics are a morass; see Doug Henwood, *After the New Economy* (New York: New Press, 2003).

18. For example, so-called private-line, or dedicated, international circuits leased by carriers to corporations, government agencies, and Internet service providers enjoyed unprecedented growth during the late 1990s. Organization for Economic Cooperation and Development, *OECD Communications Outlook 2001* (N.p.: OECD, 2001), 54 and 65 (table 3.10).

19. Via Wal-Mart's unmatched data-warehouse and transnational-network facilities, "the previous day's information, through midnight, on over 10 million customer transactions is available for every store in every country before 4 AM the following day." Prepared statement of Wal-Mart Stores, Inc., "The Role of Standards in the Growth of Global Electronic Commerce," Testimony before the Subcommittee on Science, Technology, and Space of the Committee on Commerce, Science, and Transportation, U.S. Senate, 28 October 1999, 4.

20. By 2003, 90 percent of big companies maintained toll-free customer numbers. Deborah Ball, "Toll-Free Tips: Nestle Hotlines Yield Big Ideas," WSJ, 3 September 2004, A7.

21. Steve Lohr, "The Tech Rebound That Isn't Quite," NYT, 23 June 2003, C1 (cites 60 percent); Roger Alcaly, *The New Economy* (New York: Farrar, Straus, and Giroux, 2003), 55 (cites 35 percent). James W. Cortada references an internal IBM estimate that ICT capital investment increased from 38 to 55 percent of all capital spending between 1990 and 2001. Cortada, *Digital Hand,* 376.

22. U.S. Congress, Senate, 98th Cong., 1st Sess., S. Print 98–22, Committee on Commerce, Science, and Transportation, 11 March 1983, *Long-Range Goals in International Telecommunications and Information: An Outline for U.S. Policy* (Washington, D.C.: Government Printing Office, 1983), 5.

23. For useful, brief, and up-to-date analyses of selective instances of privatization, see Jochen Boekhoff, "Telecommunications in Mexico, Uruguay, and Argentina: A Tale of Contrasts," and Thomas Thummel and Max Thummel, "Privatization of Telecommunications in Japan," in *Limits to Privatization: How to Avoid Too Much of a Good Thing, a Report to the Club of Rome,* ed. Ernest Ulrich von Weizsacker, Oran R. Young, and Matthias Finger (London: Earthscan, 2005), 72–76, and 76–79.

24. Bjorn Wellenius, Carlos Alberto Primo Braga, and Christine Zhen-Wei Qiang, "Investment and Growth of the Information Infrastructure: Summary Results of a Global Survey," *Telecommunications Policy* 24 (2000): 639, 642.

25. International Telecommunications Union, *Yearbook of Statistics: Telecom Services Chronological Time Series, 1991–2000* (Geneva: ITU, 2001).

26. U.N. Industrial Development Organization, *Industrial Development Report 2004,* xxi; retrieved 17 February 2006, www.unido.org/doc/view?document id=2603 2&language code=en. See Dicken, *Global Shift,* 27–28.

27. This figure includes profits earned in developed-country markets. Jon E. Hilsenrath, "U.S. Multinationals Reap Overseas Bounty," WSJ, 4 April 2005, A2. For transnational business reliance on networks, see Dicken, *Global Shift,* 157; John V. Langdale, "The Geography of International Business Telecommunications: The Role of Leased Networks," *Annals of the Association of American Geographers* 79 (1989): 501–22.

28. Kathryn Kranhold, "GE Pins Hopes on Emerging Markets," WSJ, 2 March 2005, A3, A10; Saritha Rai, "Chief Sees Surge in G.E. Business in India," NYT, 28 May 2005, B13. For another example, see Eric Bellman, "Nokia to Build Cellphone Plant in India to Meet Rising Demand," WSJ, 7 April 2005, B4.

29. Organization for Economic Cooperation and Development, "FDI Outflows from U.S. Hit Record USD 252 Billion In 2004," retrieved June 2005, http://www .oecd.org/document/54/0,2340,en 2649 201185 35033718 1 1 1 1,00.html.

30. U.S. Congress, *Long-Range Goals in International Telecommunications and Information,* 22.

31. I differ here from David Bollier, who asserts that a so-called knowledge commons, a species of "gift economy," sustained communitarian moral values that were subsequently undercut by market incursions. Yet Bollier goes out of his way to emphasize that the knowledge commons and the market can coexist and "invigorat[e] each other." David Bollier, *Silent Theft: The Private Plunder of Our Commonwealth* (New York: Routledge, 2002), 37.

32. Harvey, *New Imperialism,* 137–82.

33. Michael Perelman, *Steal This Idea: Intellectual Property Rights and the Corporate Confiscation of Creativity* (New York: Palgrave, 2002), 70–71.

34. Francis C. Steckel, "Cartelization of the German Chemical Industry, 1918–1925," *Journal of European Economic History* 19.2 (Fall 1990): 329–51. The American Chemical Society, which annually garners as much as $375 million in fees paid to access its Chemical Abstracts Service database, is fighting to limit free access to chemical information via PubChem, a free database established by the U.S. National Institutes

of Health. Emma Marris, "Chemistry Society Goes Head to Head with NIH in Fight over Public Database," *Nature* 435 (9 June 2005): 718–19.

35. Hugh G. J. Aitken, *The Continuous Wave: Technology and American Radio, 1900–1932* (Baltimore: Johns Hopkins University Press, 1985).

36. John Gimbel, *Science, Technology, and Reparations: Exploitation and Plunder in Postwar Germany* (Stanford, Calif.: Stanford University Press, 1990), 43.

37. This figure is important because it is about equal to the sum demanded of Germany by the Soviet Union in war reparations, which the United States opposed. Caroline Eisenberg, *Drawing the Line: The American Decision to Divide Germany, 1944–1949* (Cambridge: Cambridge University Press, 1996).

38. Herbert I. Schiller and Anita R. Schiller, "Libraries, Public Access to Information, and Commerce," in *The Political Economy of Information,* ed. Vincent Mosco and Janet Wasko (Madison: University of Wisconsin Press, 1988), 146–66, esp. 156–63; Schiller, *Who Knows,* 47–78; John E. Buschman, *Dismantling the Public Sphere* (Westport, Conn.: Libraries Unlimited, 2003); Dan Schiller, *Telematics and Government* (Norwood, N.J.: Ablex, 1982), 191–210.

39. Schiller, *Telematics and Government,* 210–15.

40. Schiller, *Digital Capitalism,* 143–202.

41. Chin-tao Wu, *Privatising Culture: Corporate Art Intervention since the 1980s* (London: Verso, 2002); Schiller, *Culture, Inc.;* Carol Vogel, "Wal-Mart Heiress Is High Bidder for a Durand Painting Sold by the New York Public Library," NYT, 13 May 2005, A19; Michael Kimmelman, "Civic Treasure: A Need for Transparency, Not Secrecy," NYT, 16 May 2005, B1, B8.

42. For the earlier campaigns, see Schiller, *Who Knows,* 58–65; Lauren Bayne Anderson, "Printing Office Monopoly Ends in a Symbolic Victory for Bush," WSJ, 6 June 2003, B6.

43. Anita R. Schiller, "Shifting Boundaries in Information," *Library Journal,* 1 April 1981, 705–9; Anita R. Schiller, "Information as a Commodity: There's No Such Thing as a Free Hunch," *Technicalities* 2.6 (June 1982): 3–5. Herbert I. Schiller and Anita R. Schiller identify "the turning point" as occurring early in the Reagan administration, with the report to the National Commission on Libraries and Information Science from the Public Sector/Private Sector Task Force in August 1981. Schiller and Schiller, "Libraries, Public Access to Information, and Commerce," 159. For a useful update of this changing area, see James A. Jacobs, James R. Jacobs, and Shinjoung Yeo, "Government Information in the Digital Age: The Once and Future Federal Depository Library Program," *Journal of Academic Librarianship* 31.3 (May 2005): 198–208.

44. Schiller, *Who Knows,* 115–34; Jurgen Scheffran, "Privatization in Outer Space: Lessons from Landsat and Beyond," in *Limits to Privatization: How to Avoid Too Much of a Good Thing, A Report to the Club of Rome,* ed. Ulrich Von Weizsacker, Oran R. Young, and Matthias Finger (London: Earthscan, 2005), 79–84; Eric Lichtblau, "U.S. to Rely More on Private Companies' Satellite Images," NYT, 13 May 2003, A26.

45. Paul Zurkowski of the Information Industry Association quoted in Schiller, *Who Knows*, 47.

46. Frank Webster, "Information: A Sceptical Account," *Advances in Librarianship* 24 (2000): 10.

47. Perelman, *Steal This Idea.*

48. W. Hamilton, "Temporary National Economic Committee Monograph No. 31: Patents and Free Enterprise," U.S. Senate Committee Print, 76th Cong., 3d Sess. (Washington, D.C.: Government Printing Office, 1941), 53. Thanks to Dan Wright for bringing this quote to my attention.

49. Ambassador Charlene Barshefsky, "The Networked World Initiative: Trade Policy Enters a New Era," Speech before the Federal Communications Bar Association, Washington, D.C., 23 October 2000, 3; retrieved 18 February 2006, http://canberra.usembassy.gov/hyper/2000/1023/epf110.htm.

50. Perelman, *Steal This Idea.* By the 2000s, the U.S. Department of Commerce boasted an undersecretary for intellectual property and at least one intellectual-property attaché. Neil King, Jr., "Sisyphus in China: U.S. Lawyer's Antipiracy Task Is Endless," WSJ, 26 July 2005, A17, A22.

51. Keith E. Maskus, *Intellectual Property Rights in the Global Economic Context* (Washington, D.C.: Institute for International Economics, 2000), 1; Barshefsky, "Networked World Initiative," 3.

52. Pradip N. Thomas, "Copyright and Emerging Knowledge Economy in India," *Economic and Political Weekly* 36.24 (16 June 2001): 2150, 2154; Susan K. Sell, *Private Power, Public Law: The Globalization of Intellectual Property Rights* (Cambridge: Cambridge University Press, 2003).

53. Paul Klebnikov, "The Twain Shall Meet," *Forbes,* 27 February 1995, 74–75.

54. Toby Miller, Nitin Govil, John McMurria, and Richard Maxwell, *Global Hollywood* (London: BFI, 2001), 207.

55. Milton Mueller, *Ruling the Root* (Cambridge: Massachusetts Institute of Technology, 2002), 231–34.

56. U.S. Patent and Trademark Office, Information Products Division Technology Assessment and Forecast Branch, "Patenting Trends Calendar Year 2001," retrieved 21 February 2006, www.uspto.gov/web/offices/ac/ido/oeip/taf/pat tr01.htm; Sabra Chartrand, "Patents," NYT, 9 February 2004, C2.

57. See official World Intellectual Property Organization statistics at http://www.wipo.int/ipstats/en/publications/a/pdf/patents.pdf.

58. U.S. National Science Board, *Science and Engineering Indicators 2004,* vol. 1, 6–23. Daniel Del Castillo, "The Arab World's Scientific Desert," *Chronicle of Higher Education,* 5 March 2004, A36.

59. Judith Miller, "Globalization Widens Rich-Poor Gap, U.N. Report Says," NYT, 13 July 1999, A8.

60. World Bank, *World Development Report 1998–99: Knowledge for Development* (New York: Oxford University Press, 1999), 27.

61. "Monti's Wrecking Crew," WSJ, 7 August 2003, A10.

62. Donald G. McNeil Jr., "India Alters Law on Drug Patents," NYT, 24 March 2005, A1, A9.

63. Bronwyn Parry, *Trading the Genome: Investigating the Commodification of Bio-Information* (New York: Columbia University Press, 2004), 258.

64. Marie Claire Leger, "Clinical Globalization: Pharmaceutical Research in the Global South" (Ph.D. dissertation, University of Illinois at Urbana-Champaign, 2004), 52.

65. Schiller, *Information and the Crisis Economy*, 12.

66. Simon London, "U.S. Launches Drive to Regain Top Spot in Supercomputing," FT, 27 May 2003, 14.

67. Andy Pasztor, "Europe Aims to Make Big Satellites," WSJ, 16 June 2005, B5.

68. "High-Tech Brain Drain" (editorial), WSJ, 5 May 2005, A14; William J. Broad, "U.S. Is Losing Its Dominance in the Sciences," NYT, 3 May 2004, A1, A19; Vinton Cerf and Harris N. Miller, "America Gasps for Breath in the R&D Marathon," WSJ, 27 July 2005, A12.

69. Dicken, *Global Shift*, 430.

70. Randolph E. Schmid, "Information a $623–Billion Industry in '97, U.S. Says," *Los Angeles Times*, 25 October 1999, C3.

71. Robert Brenner, *The Boom and the Bubble* (London: Verso, 2002). Brenner has usefully updated his analysis in Robert Brenner, "New Boom or New Bubble," *New Left Review* 25 (January–February 2004): 57–100.

72. Shirley Leung, "A Glutted Market Leaves Food Chains Hungry for Sites," WSJ, 1 October 2003, A1, A12.

73. Jon E. Hilsenrath, "Factory Employment Is Falling World-Wide," WSJ, 20 October 2003, A2, A8.

74. Makoto Ito, "The Japanese Economy in Structural Difficulties," *Monthly Review* 56.11 (April 2005): 32–44.

75. As of March 2005, Japan held $679.5 billion in U.S. Treasury bills, bonds, and notes; China held $223.5 billion; and Hong Kong held $45.2 billion. U.S. Treasury holdings by foreign entities overall nearly doubled between September 2001 and March 2005, from $992.2 billion to $1,977.1 billion. See U.S. Treasury, "Major Foreign Holders of Treasury Securities," retrieved 2 June 2005, www.treasury.gov/tic.mfh.txt; Greg Ip, "Could Overseas Financing Hurt the U.S.?" WSJ, 26 April 2004, A2; Floyd Norris, "Foreigners May Not Have Liked the War, but They Financed It," NYT, 12 September 2003, C1.

76. Christopher Swann, "Concerns for Dollar as Central Banks Sell Off Assets," FT, 17 May 2005, 1.

77. Paul A. Volcker, "An Economy on Thin Ice," *Washington Post*, 10 April 2005, B7. Also see Chris Giles, "BIS Warns Increasing Levels of Debt Are Creating Conditions for Financial Crises," FT, 28 June 2005, 1. It should be noted that over 40 percent of the U.S. current account deficit is made up of intrafirm transfers; this is not merely a

simple deficit caused by an excess of imports over exports in the traditional nation-centered sense.

78. Ellen Meiksins Wood, *Empire of Capital* (London: Verso, 2003), 137–42. Thanks to Pedro Caban for this source.

79. The Pentagon was building more than seventy major weapons systems at an overall nominal cost of $1.3 trillion as of 2005, and these relied ever more pervasively on networks and network applications. Tim Weiner, "Report Says Pentagon Spending on Weapons to Soar," NYT, 1 April 2005, C3; Jonathan Karp and Andy Pasztor, "Pentagon Work: High Tech Has High Risk," WSJ, 2 May 2005, B2; Tim Weiner, "Air Force Seeks Bush's Approval for Space Arms," NYT, 18 May 2005, A1, C4; U.S. Government Accountability Office, "Defense Acquisitions: The Global Information Grid and Challenges Facing Its Implementation," GAO-04-858, retrieved 29 July 2004, www.gao.gov.

80. Immanuel Wallerstein, "The Eagle Has Crash Landed," *Foreign Policy* 131 (July–August 2002): 60–68; Christopher Marquis, "As Bush Confers with NATO, U.S. Is Seen Losing Its Edge," NYT, 28 June 2004, A8; Giovanni Arrighi, "Hegemony Unraveling—1," *New Left Review* 32 (March–April 2005): 23–80.

81. Chalmers Johnson, *The Sorrows of Empire* (New York: Metropolitan Books, 2004).

82. Michael Perelman, *Class Warfare in the Information Age* (New York: St. Martin's Press, 1998), 80–82.

83. Joyce Jones, "Cyberwise Private Eyes," *Black Enterprise* 29.5 (December 1998): 39–40.

84. Michael Orey, "Why We Now Need a National Association for Data Destruction," WSJ, 30 January 2002, A1, A8.

85. John Markoff, "Taking Snooping Further," NYT, 25 February 2006, C1.

86. Michael Godwin, "Digital Rights Management: A Guide for Librarians," American Library Association, Office of Information Technology Policy, OITP Technology Policy Brief, 1; retrieved 12 February 2006, www.ala.org/ala/washoff/Woissues/copyrightb/digitalrights/digitalrightsmanagement.htm.

87. Marcelo Prince, "Out of Memory," WSJ, 24 October 2005, R9.

88. Brian Grow, "Hacker Hunters," *Business Week*, 30 May 2005, 74–82 ("Hacking into computer networks was long seen as little more than a prank, and punishment was typically a slap on the wrist. That's beginning to change, however" [77]).

89. Robert O'Harrow Jr., *No Place to Hide* (New York: Free Press, 2005). For a few recent instances, see Riva Richmond, "Network Giants Join Campaign to Beef Up Security of Systems," WSJ, 27 October 2004, B2; Riva Richmond, "Job of Guarding Web Is Shifting to the Network's Infrastructure," WSJ, 19 May 2005, B4; Li Yuan, "Companies Face System Attacks from Inside, Too," WSJ, 1 June 2005, B1, B4; Robert Block, "In Terrorism Fight, Government Finds a Surprising Ally: FedEx," WSJ, 26 May 2005, A1, A5; Gary Fields, "Ten-Digit Truth Check," WSJ, 7 June 2005, B1, B6;

David Pringle, "Security Woes Don't Slow Reed's Push into Data Collection," WSJ, 3 June 2005, C1, C4.

90. U.S. Government Accountability Office, "Technology Assessment: Cybersecurity for Critical Infrastructure Protection," GAO-04-321, 28 May 2004, Abstract, retrieved 3 June 2005, www.gao.gov/new.items/d04321.pdf; quote from U.S. Government Accountability Office, "Critical Infrastructure Protection: Department of Homeland Security Faces Challenges in Fulfilling Cybersecurity Responsibilities," GAO-05–434, 26 May 2005, Abstract, retrieved 3 June 2005, www.gao.gov/new.items/d05434.pdf.

91. Eric Lichtblau and James Risen, "Spy Agency Mined Vast Data Trove, Officials Report," NYT, 24 December 2005, A12.

92. Scott Shane, "Attention in N.S.A. Debate Turns to Telecom Industry," NYT, 11 February 2006, A11.

93. Schiller, *Who Knows,* xiv–xv.

94. Glen Hubbard, Paul Krugman, Lewis H. Lapham, and Peter G. Peterson, "The Iceberg Cometh: Can a Nation of Spenders Be Saved?" *Harper's,* June 2005, 45.

95. Robert E. Rubin, "'We Must Change Policy Direction,'" WSJ, 24 January 2006, A20.

96. See www.freepress.net.

97. See, for example, Ted Koppel, "Take My Privacy, Please!" NYT, 13 June 2005, A19.

98. In one mainstream tally, these issues include "controlling spam, dealing with use of the Internet for illegal purposes, resolving the digital divide between developed and developing countries; protecting intellectual property; . . . protecting privacy and freedom of expression; and facilitating and regulating e-commerce." In addition, Internet governance includes control of the domain-name system and of navigation aids (search engines and directories). National Research Council, Committee on Internet Navigation and the Domain Name System, Computer Science and Telecommunication Board, *Signposts in Cyberspace: The Domain Name System and Internet Navigation,* Prepublication copy (Washington, D.C.: National Academies Press, 2005), 5–3.

99. Hans Klein and Milton Mueller, "What to Do about ICANN: A Proposal for Structural Reform," Internet Governance Project, 5 April 2005, retrieved 5 February 2006, http://doc.syr.edu/miscarticles/IGP-ICANNReform.pdf. For a critique of the WSIS meeting, see Paula Chakravarty, "Who Speaks for the Governed?" *Economic and Political Weekly,* 21 January 2006, 250–57; retrieved 25 February 2006, www.epw.org.in/showIndex.php.

100. "Carter Adds Reprisals against Iran," *Washington Post,* 18 April 1980, A1, A29; quote from Oswald H. Ganley and Gladys D. Ganley, *To Inform or to Control? The New Communications Networks* (New York: McGraw-Hill, 1982), 46.

101. National Research Council, *Signposts in Cyberspace,* 5–46.

Chapter 4: Business Users and U.S. Telecommunications-System Development

1. Business users have not been entirely overlooked. See Timothy R. Haight, "The 'S.1' of Communications," in *Telecommunications Policy and the Citizen* (New York: Praeger, 1979), 241–66. I developed a book-length analysis of the role of business users in Dan Schiller, *Telematics and Government* (Norwood, N.J.: Ablex, 1982).

2. The larger social history of telecommunications remains remarkably obscure. However overdue, extensive and multifaceted research is now thankfully under way, including a study under preparation by the present writer.

3. Richard John, *Spreading the News: The American Postal System from Franklin to Morse* (Cambridge, Mass.: Harvard University Press, 1995).

4. Richard B. DuBoff, "Business Demand and the Development of the Telegraph in the United States, 1844–1860," *Business History Review* 54 (Winter 1980): 459–79; JoAnne Yates, *Control through Communication* (Baltimore: Johns Hopkins University Press, 1989); Alfred D. Chandler and James W. Cortada, eds., *A Nation Transformed by Information* (New York: Oxford University Press, 2000).

5. Claude Fischer, *America Calling: A Social History of the Telephone to 1940* (Berkeley: University of California Press, 1992); Dan Schiller, "Social Movement in Telecommunications: Rethinking the Public Service History of U.S. Telecommunications, 1894–1919," *Telecommunications Policy* 22.4–5 (1998): 397–408.

6. Richard B. DuBoff, "The Telegraph and the Structure of Markets in the United States, 1845–1890," *Research in Economic History* 8 (1983): 268.

7. Whereas, around 1904, AT&T was garnering 28 percent of the industry's private-wire revenue, by 1914 it had doubled its share of this market to 56 percent. Railroads were especially significant users of private-wire systems and often enjoyed access via special shared arrangements with Western Union. In 1914, four-fifths of telegraph-service private-wire leases (exclusive of railroads) were held by bankers and brokers, and an additional 9 percent of leases were held by meatpacking, iron, and steel companies. These private-wire systems were located disproportionately in the northern states, stretching from the Atlantic coast to the Missouri River. Some of the larger systems linked locations in the United States and Canada. Interstate Commerce Commission, *Decisions of the Interstate Commerce Commission of the United States May 1918 to August 1918*, vol. 50 of Interstate Commerce Commission Reports (Washington, D.C.: Government Printing Office, 1918): 734–35, 738–39.

8. "Telegraph Business West," WSJ, 30 December 1909, 6. I am grateful to Richard John for this reference.

9. "Railroads West Said to Be Unable to Get All the Telegraphers Needed," WSJ, 31 August 1909, 6. I am grateful to Richard John for this reference.

10. U.S. Congress, Senate, 60th Cong., 2d Sess., Document No. 725, Committee on Interstate Commerce, *Investigation of Western Union and Postal Telegraph-Cable Com-*

panies, 16 February 1909 (Washington, D.C.: Government Printing Office, 1909), 22. See Duboff, "Telegraph and the Structure of Markets in the United States," 268.

11. "Railroads West Said to Be Unable to Get All the Telegraphers Needed," 6.

12. The Merchants' Association of New York, *Inquiry into Telephone Service and Rates in New York City by the Merchants' Association of New York* (New York: Merchants' Association, 1905), 7, 8–9.

13. Ibid., 36.

14. Ibid., 37–38, 39.

15. David Weiman and Richard Levin, "Preying for Monopoly: The Case of Southern Bell, 1984–1912," *Journal of Political Economy* 102 (February 1994): 103–26; Kenneth Lipartito, *The Bell System and Regional Business: The Telephone in the South, 1877–1920* (Baltimore: Johns Hopkins University Press, 1989).

16. Alan Stone, *Public Service Liberalism: Telecommunications and Transitions in Public Policy* (Princeton, N.J.: Princeton University Press, 1993), 160. See also Schiller, "Social Movement in Telecommunications."

17. James Schwoch, "The American Radio Industry and International Communications Conferences, 1919–1927," *Historical Journal of Film, Radio, and Television* 7.3 (1987): 293–94.

18. Murray Edelman, *The Licensing of Radio Services in the United States, 1927 to 1947: A Study in Administrative Formulation of Policy* (Urbana: University of Illinois Press, 1950); Dallas Smythe, *The Structure and Policy of Electronic Communications* (Urbana: University of Illinois Press, 1957).

19. *Second Annual Report of the Federal Radio Commission to the Congress of the United States* (Washington, D.C.: Government Printing Office, 1928), 26.

20. Ibid., 27, 28.

21. Ibid., 33.

22. *Third Annual Report of the Federal Radio Commission to the Congress of the United States* (Washington, D.C.: Government Printing Office, 1929), 19.

23. *Second Annual Report of the Federal Radio Commission to the Congress of the United States,* 34.

24. Edelman, *Licensing of Radio Services in the United States,* 186.

25. Ibid., 171.

26. James M. Herring and Gerald C. Gross, *Telecommunications Economics and Regulation* (New York: McGraw-Hill, 1936, 56); Federal Communications Commission, *Proposed Report: Telephone Investigation* (Washington, D.C.: Government Printing Office, 1938), 419, 613.

27. Marion May Dilts, *The Telephone in a Changing World* (New York: Longmans, Green, 1941), 69, 132.

28. *By-Products Coal Co. v. Federal Radio Commission,* No. 4984, in *Third Annual Report of the Federal Radio Commission to the Congress of the United States,* 37.

29. *Intercity Radio Telegraph Co., appellant, v. Federal Radio Commission,* No. 4987,

in *Third Annual Report of the Federal Radio Commission to the Congress of the United States*, 40.

30. Grayson M.-P. Murphy to Honorable Clarence C. Dill, 17 March 1934, in U.S. Congress, Senate, 73d Cong., 2d Sess., *Hearings before the Committee on Interstate Commerce on S. 2910, Federal Communications Commission, March 9, 10, 13, 15, 1934* (Washington, D.C.: Government Printing Office, 1934), 217–18 (see also 142–53, 173–76).

31. The New Deal settlement in telecommunications is a major focus of my work in progress, tentatively titled "The Hidden History of U.S. Telecommunications."

32. John M. Schacht, *The Making of Telephone Unionism, 1920–1947* (New Brunswick, N.J.: Rutgers University Press, 1985).

33. This development was noticed by a pioneer in communications scholarship, perhaps because he had served during this period as chief economist at the FCC. See Smythe, *Structure and Policy of Electronic Communications.*

34. Joint Technical Advisory Committee, Institute of Radio Engineers and Radio-Television Manufacturers Assocation, *Radio Spectrum Conservation: A Program of Conservation Based on Present Uses and Future Needs* (New York: McGraw-Hill, 1952), 13.

35. On RCA, see Margaret Graham, *The Business of Research: RCA and the Videodisc* (New York: Cambridge University Press, 1986).

36. "I.B.M. Trust Suit Ended by Decree; Machines Freed," NYT, 26 January 1956, 1, 18.

37. Emerson W. Pugh and William Aspray, "Creating the Computer Industry," *IEEE Annals of the History of Computing* 18.2 (1996): 14.

38. Ibid., 9, 10 (quote).

39. James W. Cortada, "Commercial Applications of the Digital Computer in American Corporations, 1945–1995," *IEEE Annals of the History of Computing* 18.2 (1996): 20–21. For much more extended treatment, see James W. Cortada, *The Digital Hand* (New York: Oxford: University Press, 2004).

40. Pugh and Aspray, "Creating the Computer Industry," 11.

41. Duncan G. Copeland, Richard O. Mason, and James L. McKenney, "Sabre: The Development of Information-Based Competence and Execution of Information-Based Competition," *IEEE Annals of the History of Computing* 17.3 (1995): 30–56.

42. Paul Baran, "The Coming Computer Utility—Laissez-Faire, Licensing, or Regulation?" Santa Monica: RAND Corporation, P-3466, April 1967: 12–13.

43. Jason Oxman, "The FCC and the Unregulation of the Internet," Office of Plans and Policy, FCC, OPP Working Paper No. 31 (July 1999), 6.

44. The following paragraphs draw on material published in Dan Schiller, "Business Users and the Telecommunications Network," *Journal of Communication* 32.4 (1982): 84–96; and Schiller, *Telematics and Government.*

45. Federal Communications Commission, "In the Matter of Allocation of Fre-

quencies in the Bands Above 890 Mc.," Docket 11866, Comment of the Automobile Manufacturers Assocation, 15 March 1957, 849–51.

46. Business users generated 58 percent of domestic long-distance revenues in 1976, and the largest 3.9 percent of their number accounted for over three-fifths of this total. U.S. House of Representatives, 97th Cong., 1st Sess., Committee on Energy and Commerce, *Telecommunications in Transition: The Status of Competition in the Telecommunications Industry,* A Report by the Majority Staff of the Subcommittee on Telecommunications, Consumer Protection, and Finance, Committee Print 9 7–V (Washington, D.C.: Government Printing Office, 1981), 87–89.

47. In the longer work in progress on which this chapter draws, I hope to devote attention to a succession of popular-cultural projects that concurrently heaped scorn on AT&T, the core of the regulated network manager system.

48. President's Task Force on Communications Policy, *Final Report* (Washington, D.C.: Government Printing Office, 1968), chap. 1, 6. Documents detailing the deliberations of the Task Force are housed at the Lyndon Baines Johnson Presidential Library in Austin, Texas (hereafter LBJ Presidential Library).

49. "Memorandum for the President from the Chairman of the President's Task Force on Communications Policy," 2 December 1968, p. 3, in National Security File, Subject File, Communications Policy—Task Force on, LBJ Presidential Library.

50. Nicholas Johnson, "The Public Interest and Public Broadcasting: Looking at Communications as a Whole," Remarks Prepared for Delivery to the Resources for the Future and the Brookings Institution Conference on the Use and Regulation of the Radio Spectrum, Airlie House, Warrenton, Va., September 11, 1967, 1, 3; "Memorandum for the President from the Chairman of the President's Task Force on Communications Policy," 2 December 1968, p. 3, both in National Security File, Subject File, Communications Policy—Task Force on, LBJ Presidential Library.

51. "Memorandum for the President from the Chairman of the President's Task Force on Communications Policy," 2 December 1968, p. 7, in National Security File, Subject File, Communications Policy—Task Force on, LBJ Presidential Library.

52. President's Task Force on Communications Policy, *Final Report,* chap. 1, 17, 20.

53. Ibid., chap. 1, 21; chap. 6, 26–28 and 29–36.

54. Ibid., chap. 1, 9; chap. 6, 9–25.

55. Ibid., chap. 6, 16–17.

56. Alan Stone, *Wrong Number: The Breakup of AT&T* (New York: Basic Books, 1989), 338.

57. See Daniel R. Headrick, *Tools of Empire* (New York: Oxford University Press, 1981).

58. Daniel R. Headrick, *The Invisible Weapon: Telecommunications and International Politics, 1851–1945* (New York: Oxford University Press, 1991).

59. U.S. Department of Commerce, National Telecommunications and Information Administration, "Long-Range Goals in International Telecommunications and

Information: An Outline for United States Policy," 98th Cong., 1st Sess., Senate Print 98–22, 11 March 1983, 114.

60. Dan Schiller, "Comment perpétuer la domination sur les télécommunications?" *Le Monde diplomatique,* February 1985, 14–15.

61. Seth Schiesel, "FCC to Open Phone Market to Foreigners," NYT, 25 November 1997, C1, C2. See "Rules and Policies on Foreign Participation in the U.S. Telecommunications Market," International Bureau Docket Nos. 97–142 and 95–22, Report and Order, 11 Fcc Rcd 3873 (1997).

62. Andrew Ross Sorkin and Simon Romero, "Deutsche Telekom to Pay $50 Billion for U.S. Company," NYT, 24 July 2000, A1, A8.

63. In Fall 2001, around twenty-four million U.S. inhabitants possessed T1 or higher-speed Internet access lines in their offices, while fewer than 9 million enjoyed broadband, or high-speed, connections at home; thus the cultural habit of Net access, now including broadband access, was nurtured at work and only secondarily at home. Melinda Patterson Grenier, "Record Number of Office Workers Used Web Broadcasts Last Month," WSJ, 15 October 2001, B8.

64. Urs von Burg, *The Triumph of Ethernet: Technological Communities and the Battle for the LAN Standard* (Stanford, Calif.: Stanford University Press, 2001).

65. Eli M. Noam and Aine NiShuilleabhain, Introduction to *Private Networks, Public Objectives,* ed. Eli M. Noam and Aine Nishuilleabhain (Amsterdam: Elsevier, 1996), xvii.

66. U.S. Congress, Office of Technology Assessment, *Critical Connections: Communications for the Future,* OTA-CIT-407 (Washington, D.C.: Government Printing Office, 1990), 137.

67. Burg, *Triumph of Ethernet,* xiii.

68. Intranets are intraorganizational networks that make use of Internet technologies: routers and TCP/IP, as well as, typically, the HTTP application, often with additional applications such as email and file-transfer protocols.

69. Vinton Cerf, "Transforming Impact of Internet," Paper presented at the Internet and the Global Political Economy Conference, University of Washington Center for Internet Studies, Seattle, 23 September 1999.

70. Kim Cross, "B-to-B, by the Numbers," *Business 2.0* (September 1999): 109. See also Dan Schiller, "World Communications in Today's Age of Capital," *Emergences* 11.1 (2001): 59.

Chapter 5: The Crisis in Telecommunications

1. Bjorn Wellenius, Carlos Alberto Primo Braga, and Christine Zhen-Wei Qiang, "Investment and Growth of the Information Infrastructure: Summary Results of a Global Survey," *Telecommunications Policy* 24 (2000): 639 and 642.

2. International Telecommunication Union, *Yearbook of Statistics: Telecommunication Services Chronological Time Series, 1991–2000* (Geneva: ITU, 2001).

3. John Schwartz, "Text Messaging Pushed for Use as Disaster Warning Systems," NYT, 31 December 2004, A12.

4. International Telecommunication Union, *Yearbook of Statistics;* International Telecommunication Union, *World Telecommunication Development Report* (Geneva: ITU, 2002), 13.

5. U.S. Department of Commerce, National Telecommunications and Information Administration and Economics and Statistics Administration, *A Nation Online: How Americans Are Expanding Their Use of the Internet* (Washington, D.C.: Government Printing Office,2002), 9; Brian Hammond, "Commerce Cites 'Rapid Growth' in Broadband Use," *TR Daily,* 5 February 2002, 7.

6. U.S. Department of Commerce, NTIA and ESA, *A Nation Online,* 9.

7. Adam Cohen, "PayPal and Other Post-Bubble Signs of Life on the Internet," NYT, 6 February 2002, A26 (Cohen asserts that twenty-one million wired computers nationwide had installed photo-editing and -sharing software).

8. Paul Tayloir and Aline Van Duyn, "Head to Head: Cable and Phone Groups Make a 'Triple Play' for U.S. Households," FT, 12 January 2005, 11.

9. George Gilder, *Telecosm: How Infinite Bandwidth Will Revolutionize Our World* (New York: Free Press, 2000), 3.

10. Antonio Pasquali, "The Information Society: A Case for Setting Up an International Tribunal," *Media Development* 4 (2002): 3.

11. David Bank, "Cisco Profit Surges on Cost Controls," WSJ, 7 August 2002, A3; Scott Thurm, "A Go-Go Giant of Internet Age, Cisco Is Learning to Go Slow," WSJ, 7 May 2003: A1, A7.

12. Written Statement of Michael K. Powell, chairman of the FCC, on "Health of the Telecommunications Sector: A Perspective from the Commissioners of the FCC," Before the Subcommittee on Telecommunications and the Internet, Committee on Energy and Commerce, U.S. House of Representatives, 26 February 2003, 4; retrieved 5 February 2006, http://hraunfoss.fcc.gov/edocs public/attachmatch/DOC-231535A1 .pdf.

13. Paul Klemperer, "The Wrong Culprit for Telecom Trouble," FT, 26 November 2002, 15; Eli Noam, "Too Weak to Compete," FT, 19 July 2002, 11 (quote).

14. "France Telecom Posts Big Loss," WSJ, 6 March 2003, B4.

15. Silvia Ascarelli and Almar Latour, "Europe's Reckoning May Not Be Over Just Yet," WSJ, 11 March 2003, C1.

16. Ken Belson, "Japan's Cellphone Giant Casts a Paler Shadow," NYT, 2 December 2002, C5.

17. Mike Esterl, "Deutsche Telekom to Cut 19,000 Jobs," WSJ, 3 November 2005, A3.

18. Dan Schiller, *Telematics and Government* (Norwood, N.J.: Ablex, 1982). See also James Cortada, *The Digital Hand* (New York: Oxford University Press, 2004), for a valuable though uncritical study of the growth of computerization—which soon took in networking—within U.S. manufacturing.

19. International Telecommunications Union, *World Telecommunication Development Report*, A56–59, table 14.

20. Ibid., 41.

21. Howard Gleckman, John Carey, Russell Mitchell, Tim Smart, and Chris Roush, "The Technology Payoff," *Business Week*, 14 June 1993, 57–68.

22. For one informative example, see Seth Schiesel, "The No. 1 Customer: Sorry, It Isn't You," NYT, 23 November 1997, B1. For academic analyses, see S. D. Oliner and D. E. Sichel, "The Resurgence of Growth in the Late 1990s: Is Information Technology the Story?" Working Paper, May 2000; retrieved 5 February 2006, http://www.federal-reserve.gov/pubs/feds/2000/200020pap.pdf; Wellenius, Primo Braga, and Zhen-Wei Qiang, "Investment and Growth of the Information Infrastructure," 639, 642.

23. Almar Latour and David Pringle, "Vodafone Changes CEOs at Key Time," WSJ, 19 December 2002, A3. By 2004, Vodafone's units spanned twenty-six countries, but it still possessed nothing approaching 100 percent ownership of many of its affiliates. Vodafone's stakes varied from full (100 percent) ownership in the case of its parent Vodafone UK, Vodafone (Germany), and Vodafone (Spain), to 3 percent in China Mobile; 44 percent in SFR (France); 35 percent in Vodacom (South Africa); 45 percent in Verizon Wireless (United States). David Reiley and David Pringle, "Vodafone Is Challenge for Investors Tracking State of Global Empire," WSJ, 2 September 2004, C1, C3.

24. Jo Johnson, "France Telecom Sells Dutch Cable Unit," FT, 27 December 2002, 11.

25. Ken Belson, "Japan's Cellphone Giant Casts a Paler Shadow," NYT, 2 December 2002, C5.

26. This quote, which only repeats figures in wider circulation, is from Gilder, *Telecosm*, 11.

27. Dan Schiller, *Bad Deal of the Century: The Worrisome Implications of the World-Com-MCI Merger* (Washington, D.C.: Economic Policy Institute, 1998), 2. Odlyzko is mentioned in Shawn Young, "Why the Glut in Fiber Lines Remains Huge," WSJ, 12 May 2005, B1, B10.

28. Dan Schiller, *Digital Capitalism: Networking the Global Market System* (Cambridge: Massachusetts Institute of Technology Press, 1999), 68.

29. Shawn Young, Deborah Solomon, and Dennis K. Berman, "Qwest Engaged in Fraud, SEC Says," WSJ, 22 October 2004, A3.

30. Rebecca Blumenstein, Joann S. Lublin, and Shawn Young, "Sprint Forced Out Top Executives over Questionable Tax Shelter," WSJ, 5 February 2003, A1, A8; Rebecca Blumenstein and Carol Hymowitz, "Inside the Tough Call at Sprint: Fire Auditor or Top Executives?" WSJ, 10 February 2003, A1, A6; Jason Singer and Shawn Young, "Ousted Executives of Sprint Hired by Japan Telecom," WSJ, 22 September 2003, A3.

31. Peter Grant, "On-Demand TV Expands via Underused Fiber Highways," WSJ, 17 December 2004, B1; Young, "Why the Glut in Fiber Lines Remains Huge," B1, B10 (quote).

32. TeleGeography, Inc., *International Bandwidth 2001* (Washington, D.C.: TeleGeography, Inc., 2001), 23.

33. Rob Frieden, "The Telecommunications Meltdown in the United States: Reasons and Solutions," 31 n.24, retrieved 18 February 2006, www.intel.si.umich.edu/tprc/papers/2003/178/Telecom Meltdown.pdf.

34. Federal Communications Commission, International Bureau Report, "Trends in the International Telecommunications Industry," September 2005, table 4.

35. In the late 1990s, households with incomes of seventy-five thousand dollars and higher were more than twenty times more likely to have access to the Internet than those at the lowest income levels, and more than nine times as likely to have a computer at home. U.S. Department of Commerce, National Telecommunications and Information Administration, *Falling through the Net: Defining the Digital Divide; A Report on the Telecommunications and Information Technology Gap in America* (Washington, D.C.: Government Printing Office, 1999), xii–xiii. These gaps have since narrowed.

36. Anne Marie Squeo, "Rural-Phone Subsidy's Shortfall Could Be Costly for Consumers," *WSJ*, 1 November 2004, A2.

37. TeleGeography, *International Bandwidth 2001*, 124.

38. Pippa Norris, *Digital Divide: Civic Engagement, Information Poverty, and the Internet Worldwide* (New York: Cambridge University Press, 2001), 45.

39. Judith Miller, "Globalization Widens Rich-Poor Gap, U.N. Report Says," *NYT*, 13 July 1999, A8.

40. World Association for Christian Communication, "Media Ownership and Citizen Access: A Global Overview," in *Consolidated Report of WACC's Media Ownership Programme, 1997–2000* (London: World Association for Christian Communication, 2000), 2.

41. "Indonesia Rescinds Rise in Phone Rates," *NYT*, 16 January 2003, W7.

42. Judith Mariscal, Carla Bonina, and Julio Luna, "New Market Scenarios in Latin America," in *Digital Poverty: Latin American and Caribbean Perspectives*, ed. Hernan Galperin and Judith Mariscal (Lima: Regional Dialogue on the Information Society, 2005), 66; retrieved 1 February 2006, www.dirsinet/index.php?newlang=eng.

43. Juan Forero, "Latin America Fails to Deliver on Basic Needs," *NYT*, 22 February 2005, A1, C2. For a more detailed study of the Latin American context, see Sybil Rhodes, *Social Movements and Free-Market Capitalism in Latin America: Telecommunications Privatization and the Rise of Consumer Protest* (Albany: State University of New York Press, 2006).

44. Adam Aljewidz, "Telkom's IPO Makes It, Struggling," *WSJ*, 5 March 2003, C5.

45. U.S. House of Representatives, House Committee on Energy and Commerce, Subcommittee on Telecommunications and the Internet, Statement of Commissioner Kathleen Q. Abernathy, "Health of the Telecommunications Sector," 26 February 2003, 3; retrieved 25 February 2006, http://energycommerce.house.gov/108/hearings/02262003Hearing781/Abernathy1271/htm.

46. Shawn Young, "Less May Be More," WSJ, 25 October 2004, R10; Ken Belson, "Market Place," NYT, 25 October 2004, C4; Sara Silver, "Lucent's Profit Crutch—Pensions," WSJ, 28 December 2005, C1, C4.

47. Mark Heinzl, "Nortel Again Delays Financial Report," WSJ, 12 November 2004, A3.

48. Ken Brown and Mark Heinzl, "Nortel Board Finds Accounting Tricks behind '03 Profits," WSJ, 2 July 2004, A1, A8.

49. Dennis K. Berman, "Corning Tries to Adapt to Changing Times," WSJ, 6 March 2003, B4.

50. Gilder, *Telecosm*, 210; David R. Baker, "JDS Uniphase Set to Slash Workforce," *San Francisco Chronicle*, 25 October 2002, D1.

51. Mark Maremont, "Tyco Posts Loss, Will Cut 7,200 Jobs," WSJ, 5 November 2003, A3, A8.

52. Dennis K. Berman, "Bush Is Expected to Approve Global Crossing Deal," WSJ, 9 September 2003, A2.

53. New Paradigm Resources Group, Inc., *Measuring the Economic Impact of the Telecommunications Act of 1996: Telecommunications Capital Expenditures (1996–2001)*, Prepared for Competitive Telecommunications Association (Chicago: New Paradigm Resources Group, 2002).

54. Larry F. Darby, Jeffrey A. Eisenach, and Joseph S. Kraemer, "The CLEC Experiment: Anatomy of a Meltdown," Progress on Point Release 9.23 (Washington, D.C.: Progress and Freedom Foundation, 2002), 5.

55. Written Statement of Michael K. Powell, 26 February 2003, 6, 8.

56. Steven Jackson, "Ex-Communication: Competition and Collusion in the American Prison Telephone Industry," *Critical Studies in Media and Communications* 22.4 (October 2005): 263–80; see also the Campaign to Promote Equitable Telephone Charges Web page, www.curenational.org/etc/about the campaign.htm.

57. Matt Richtel, "Unions Struggle as Communications Industry Shifts," NYT, 1 June 2005, C1, C4.; Gary Fields, Timothy Aeppel, Kris Maher, and Janet Adamy, "Reinventing the Union," WSJ, 27 July 2005, B1, B2; Matt Richtel, "In Wireless World, Cingular Bucks the Antiunion Trend," NYT, 21 February 2006, C1, C2.

58. Alicia Chang, "Report: Verizon Service Quality in New York Declining," Associated Press, 7 May 2003; Fred O. Williams, "PSC finds 'Alarming' Problems with Verizon service," *Buffalo News*, retrieved 23 May 2003, www.buffalonews.com/editorial/20030523/1000921.asp.

59. Statement of Michael J. Copps, FCC commissioner, Before House Subcommittee on Telecoms and the Internet, Committee on Energy and Commerce, U.S. House of Representatives, 26 February 2003, 6; retrieved 5 February 2006, http://hraunfoss.fcc.gov/edocs public/attachmatch/DOC-231559A1.pdf.

60. "Joint Venture with PCCW Ltd. Is Being Written Down Sharply," WSJ, 24 February 2003, B3.

61. Kevin J. Delaney, "Alcatel Posts Loss, Warns on Sales," WSJ, 5 February 2003, B3.

62. Written statement of Michael K. Powell, 26 February 2003, 5; Ken Brown and Almar Latour, "Phone Industry Faces Upheaval as Ways of Calling Change Fast," WSJ, 25 August 2004, A1, A8.

63. Dennis K. Berman, "Bride or Bridesmaid? AT&T and MCI May Compete for Suitors," WSJ, 2 August 2004, C1.

64. David Pringle, "Slower Growth Hits Cellphone Services Overseas," WSJ, 23 May 2005, A1, A6.

65. Shawn Young, "AT&T Posts $7.11 Billion Loss in Wake of Charges," WSJ, 22 October 2004, B3; John Seward and Almar Latour, "MCI Plans $3.5 Billion Charge to Reflect Network-Value Drop," WSJ, 19 October 2004, B2; Christine Nuzum, "Sprint Intends to Write Down Long-Distance Assets, Cut Jobs," WSJ, 18 October 2004, B4.

66. Ken Belson, "AT&T Won't Seek New Residential Customers," NYT, 23 July 2004, A1, C3; Ken Brown and Almar Latour, "Phone Industry Faces Upheaval as Ways of Calling Change Fast," WSJ, 25 August 2004, A1, A8; Shawn Young, "AT&T Posts 80 Percent Drop in Net, Confirms Consumer Retreat," WSJ, 23 July 2004, A11.

67. Committee on the Internet and the Evolving Information Infrastructure, Computer Science and Telecommunications Board, National Research Council, *The Internet's Coming of Age* (Washington, D.C.: National Academy Press, 2001), 151–76.

68. The FCC chair estimated this debt worldwide at "nearly $1 trillion" at the end of February 2003. Written Statement of Michael K. Powell, 26 February 2003, 4. The overall amount of corporate debt classed as in default or in "distress" globally increased 38 percent to $942 billion in 2002, according to Jonathan Stempel, "Troubled Corporate Debt Swells to $942 Billion," Reuters, 7 February 2003, retrieved 11 February 2003, http://biz.yahoo.com/rf/030207/financial distressed debt 1.html.

69. European Central Bank, "EU Banking Sector Stability," February 2003, 16; retrieved 3 March 2003, www.ecb.int/pub/pdf/other/eubksectorstabilityeu.pdf. See also "ECB Tries to Allay Fears of an EU Bank Crisis," WSJ, 25 February 2003, A13.

70. Erik Portanger, "Telecom Debt Collapses on Surprised Buyers," WSJ, 15 January 2003, C1, C11. In the United States, some state pension fund managers, perhaps anxious that the political heat under them might intensify, sought redress. The Illinois State Universities Retirement System, for example, burned by having lost around $30 million on a 2001 WorldCom bond sale, took legal action against the carrier and the accountants and underwriting investment banks that had managed the offering. "SURS Takes Action in Securities Litigation," *The Advocate* 12.2 (2003): 1, 4.

71. Remarks of Michael K. Powell at the Goldman Sachs Communicopia XI Conference, 2 October 2002, New York; retrieved 6 February 2006, http://hraunfoss.fcc .gov/edocs public/attachmatch/DOC-22629A1.pdf.

72. Andrew Hill, "Buffett Likens Derivatives to Weapons of Mass Destruction," FT, 4 March 2003, 1.

73. In late 2004, the Securities and Exchange Commission decided not to file civil charges against the former chairman of Global Crossing, Gary Winnick. Bernard Ebbers, the CEO of WorldCom, was convicted of fraud in 2005 and sentenced to twenty-five years in prison; he has appealed. The CEO of Tyco International, L. Dennis Kozlowski, was convicted in 2005 on fraud, conspiracy, and grand larceny charges. Qwest CEO Joseph Naccio was charged with illegal insider trading in December 2005. Deborah Solomon, "SEC Won't Charge, Fine Global Crossing Chairman," WSJ, 13 December 2004, A1, A13; Almar Latour, Shawn Young, and Li Yuan, "Ebbers Is Convicted in Massive Fraud," WSJ, 16 March 2005, A1, A10; Ken Belson, "WorldCom Head Is Given 25 Years for Huge Fraud," NYT, 14 July 2005, A1, C4; Andrew Ross Sorkin, "Ex-Chief and Aide Guilty of Looting Millions at Tyco," NYT, 18 June 2005, A1, B4; Greg Griffin, "Naccio Hit with 42 Insider-Trading Counts," Denver Post, 21 December 2005, 1A, 14A.

74. Almar Latour, "SBC Gets Big Internet-Phone Pact," WSJ, 21 September 2004, B8.

75. Statement of Michael J. Copps, 26 February 2003, 2.

76. Ken Belson, "Ex-Chief of WorldCom Is Found Guilty in $11 Billion Fraud," NYT, 16 March 2005, A1, C8.

77. Om Malik, *Broadbandits: Inside the $750 Billion Telecom Heist* (New York: John Wiley and Sons, 2003), 292.

78. Hal R. Varian, "Economic Scene," NYT, 14 December 2000, C2.

79. For some indications, see Peter Grant and Dionne Searcey, "Verizon, SBC Take TV Battle to Statehouses," WSJ, 17 May 2005, B1, B3.

80. "Bush Sees Broadband Access as a U.S. Goal," WSJ, 29 March 2004, A4; Anne Marie Squeo, "Election Pledge: Broadband Access for All," WSJ, 14 September 2004, A4.

81. Jesse Drucker and Christopher Rhoads, "Phone Consolidation May Cost Corporate Clients Clout," WSJ, 4 May 2005, B1, B3; Matt Richtel and Ken Belson, "Phone Mergers Seen as a Curb on Price Wars," NYT, 15 February 2005, A1, C6.

82. "The Antitrust Wars" (editorial), WSJ, 11 May 2005, A14.

83. Paul Kirby, "NSTAC to Ponder Recommendations for Improving Response to Disasters," TR Daily, 13 October 2005, 1–2; James Risen and Eric Lichtblau, "Bush Lets U.S. Spy on Callers without Courts," NYT, 16 December 2005, A1, A22; Eric Lichtblau and James Risen, "Spy Agency Mined Vast Data Trove, Officials Report," NYT, 24 December 2005, A1, A12; "AT&T Is Accused of Role in Eavesdropping," NYT, 1 February 2006, A14.

84. Christopher Rhoads and Amy Schatz, "Power Outages Hamstring Most Emergency Communications," WSJ, 1 September 2005, A7; Dionne Searcey and Jesse Drucker, "Phone Networks Fail Once Again in a Disaster," WSJ, 6 September 2005, A19, A21.

85. Amy Levine, legislative counsel to Rep. Rick Boucher (D., Va.), quoted in Lynn

Stanton, "Hill Staffers Hold Out Hope for Telecom Bill This Year," *TR Daily*, 1 June 2005, 1–2.

86. Jesse Drucker and Li Yuan, "Phone Giants Are Lobbying Hard to Block Towns' Wireless Plans," WSJ, 23 June 2005, A1, A8.

87. Shawn Young, "MCI to Emerge from Bankruptcy," WSJ, 20 April 2004, B5; Peter Grant and Almar Latour, "Battered Telecoms Face New Challenge: Internet Calling," WSJ, 9 October 2003, A1, A9; Almar Latour and Kevin J. Delaney, "KaZaA Creators Connect to Phones," WSJ, 19 September 2003, B5; Almar Latour and Peter Grant, "PC Users Can Now Make Long-Distance Calls Free," WSJ, 9 October 2003, D2.

88. U.S. Supreme Court, *National Cable and Telecommunications Association et al. v. Brand X Internet Services et al.*, 27 June 2005, no. 04–277 (slip opinion); Amy Schatz, "FCC May Set Rules Allowing Bells Exclusive Access over DSL," WSJ, 3 August 2005, A4.

89. Carlta Vitzhum and Almar Latour, "Telefonica to Acquire Assets of BellSouth in Latin America," WSJ, 8 March 2004, B4; James Politi, Paul Taylor, and Leslie Crawford, "Bell South in Talks on $6bn Sale to Telefonica," FT, 5 March 2004, 1; "Telefonica," FT, 17 May 2005, 16.

90. John Authors, "Rich Pickings Send Slim into Latin American Telecoms," FT, 7 September 2004, 16.

91. H. Sender, "Asia Global Crossing Bid Signals Chinese Arrival as Merger Players," WSJ, 19 November 2002, C5; H. A. Bolande and D. Berman, "China Netcom Group to Buy Network," WSJ, 18 November 2002, A3; J. Leahy, "Saved Asia Netcom Aims for Break-Even," FT, 12 March 2003, 18; J. L. Schenker, "Asians Rush to Buy Assets of Ailing Giants: U.S. Telecom Pain Is World's Gain," *International Herald-Tribune*, 25 August 2004, retrieved 6 February 2006, http://www.iht.com/articles/535472.htm; "Slim Pickings," *The Economist*, 20 March 2004, 64.

92. Mark Maremont, "Tyco Will Sell Fiber-Optic Unit for $130 Million," WSJ, 2 November 2004, A10.

Chapter 6: The Culture Industry

1. Edward D. Horowitz, "The Ascent of Content," in *The Future of the Electronic Marketplace*, ed. Derek Leebaert (Cambridge: Massachusetts Institute of Technology Press, 1998), 93.

2. William Leach, *Land of Desire: Merchants, Power, and the Rise of a New American Culture* (New York: Pantheon, 1993).

3. Eric Hobsbawm, *The Age of Extremes: A History of the World, 1914–1995* (New York: Pantheon, 1994), 520.

4. Ithiel de Sola Pool, *Technologies of Freedom* (Cambridge, Mass.: Belknap Press, 1983), 23, 27, 28.

5. U.S. Congress, House of Representatives, 108th Cong., 2d Sess., Subcommittee on Telecommunications and the Internet, Committee on Energy and Commerce,

Hearing on "Competition in the Communications Marketplace: How Convergence Is Blurring the Lines between Voice, Video, and Data Services," 19 May 2004, serial No. 108–85 (Washington, D.C.: Government Printing Office, 2004). See also Jonathan E. Nuechterlein and Philip J. Weiser, *Digital Crossroads: American Telecommunications Policy in the Internet Age* (Cambridge: Massachusetts Institute of Technology Press, 2005), 23–30.

6. Pool, *Technologies of Freedom*, 5.

7. David A. Mindell, *Between Human and Machine: Feedback, Control, and Computing before Cybernetics* (Baltimore: Johns Hopkins University Press, 2002). For an earlier argument that convergence is an old trend, see Dwayne Winseck, *Reconvergence: A Political Economy of Telecommunications in Canada* (Creskill, N.J.: Hampton Press, 1998).

8. Frank Baldwin Jewett, "Electrical Communication: Past, Present, Future: Speech to National Academy of Sciences," *Bell Telephone Quarterly* 14 (1935): 170.

9. Ibid.

10. Ibid., 172, 171. A recent analyst observes that "[t]he radio sector, appearing in the 1920s, evolved from the learning acquired in the initial commercializing of modern electrical and telephone equipment in the 1890s." Alfred D. Chandler Jr., *Inventing the Electronic Century: The Epic Story of the Consumer Electronics and Computer Industries* (Cambridge, Mass.: Harvard University Press, 2005), 13.

11. Lloyd Espenschied, "The Origins and Development of Radiotelephony," *Proceedings of the Institute of Radio Engineers* 25.9 (September 1937): 1121.

12. Jewett, "Electrical Communication," 198.

13. Mindell, *Between Human and Machine*, 320.

14. Vertical integration is investment in businesses backward or forward in a given industry's production and distribution chain. Examples include a magazine publisher's investment in paper making, or a telephone carrier's investment in the manufacture of handsets and network facilities. Economies of scope are achieved by lowering overall unit costs as a result of investing in production of related products, which share overlapping needs for design, production, and marketing. Chandler, *Inventing the Electronic Century*, xiv.

15. Jewett, "Electrical Communication," 171.

16. This point comes through in an unusually perceptive and careful analysis of convergence: Nicholas Garnham, "Constraints on Multimedia Convergence," in *Information and Communication Technologies: Visions and Realities*, ed. William H. Dutton (New York: Oxford University Press, 1996), 108.

17. A vital and neglected nexus for this claim are the seventy-five-odd boxes of material collected by the FCC in its "telephone inquiry" of 1935–39, housed at the National Archives and Records Administration.

18. See Garnham, "Constraints on Multimedia Convergence," 116–17.

19. Jim Davis and Michael Stack, "The Digital Advantage," in *Cutting Edge: Tech-*

nology, Information Capitalism, and Social Revolution, ed. Jim Davis, Thomas Hirschl, and Michael Stack (London: Verso, 1997), 128.

20. Ibid., 128, 124–25.

21. Council of Economic Advisors, *Economic Report of the President 2005* (Washington, D.C.: Government Printing Office, 2005), 145, 150.

22. Patricia Aufderheide, *Communications Policy and the Public Interest: The Telecommunications Act of 1996* (New York: Guilford Press, 1999), 74.

23. Adam Thierer, "Telecommunications, Broadband, and Media Policy," in *Cato Handbook on Policy,* ed. Edward H. Crane and David Boaz (Washington, D.C.: Cato Institute, 2005), 400–401.

24. Jonathan E. Nuechterlein and Philip J. Weiser, *Digital Crossroads: American Telecommunications Policy in the Internet Age* (Cambridge: Massachusetts Institute of Technology Press, 2005), 27.

25. In one apt formulation, "[T]here is an imperative to re-inflect notions of the public airwaves into our digital futures." Toby Miller, Nitin Govil, John McMurria, and Richard Maxwell, *Global Hollywood* (London: BFI, 2001), 109 and 200–201.

26. Thierer, "Telecommunications, Broadband, and Media Policy," 400–401.

27. U.S. Census Bureau, *Statistical Abstract of the United States 2004–2005* (Washington, D.C.: Government Printing Office, 2004), 716 (table 1119: "Media Usage and Consumer Spending: 1999 to 2007").

28. Aufderheide, *Communications Policy and the Public Interest,* 91; Diane Mermigas, "How Companies Deal with the Net," *Electronic Media,* 19 April 1999, 34.

29. Robert W. McChesney and Dan Schiller, "The Political Economy of International Communications: Foundations for the Emerging Global Debate about Media Ownership and Regulation," United Nations Research Institute for Social Development, Technology, Business, and Society, Programme Paper No. 11, July 2003, 16–17; Diane Mermigas, "Merger Frenzy Will Continue in New Year," *Television Week,* 31 December 2001; Diane Mermigas, "Expect a Flood of Mergers in 2004," *Television-Week,* 5 January 2004.

30. Diane Mermigas, "More Work Ahead for TW's Parsons," *Television Week,* 9 February 2004, 12–13.

31. Doug Halonen, "Broadcasters Oppose Coalition Proposal," *Electronic Media,* 6 January 2003, 2.

32. Tim Burt and Joshua Chaffin, "Universal Ambitions Spur NBC," FT, 21 June 2005, 15.

33. Cate Doty, "Love before Music," NYT, 14 March 2005, C7.

34. Mermigas, "Expect a Flood of Mergers in 2004."

35. Herbert A. Lottman, "Books and the Information Future: Two European Views," *Publishers Weekly,* 4 September 1995, 15.

36. Kenneth Li, "Publish Online or Perish?" *Industry Standard,* 3 April 2000, 60–61.

37. John F. Baker, "RH Sets Up West Coast Div. for Kids' Properties," *Publishers Weekly*, 28 August 1995, 25.

38. Paul Klebnikov, "The Twain Shall Meet," *Forbes*, 27 February 1995, 74–75.

39. Melinda Fulmer, "Food Makers Cashing In by Turning Brands into Books, Toys," *Los Angeles Times*, 26 March 2000, C4.

40. Ethan Smith, "EMI, Sony BMG Set License Accord for Music Catalogs," WSJ, 20 December 2004, B5.

41. Klebnikov, "The Twain Shall Meet," 74–75.

42. Merissa Marr and Dennis K. Berman, "Behind MGM Sale: A Gamble That DVDs Will Keep Booming," WSJ, 15 September 2004, B1, B2.

43. U.S. National Research Council, Committee on Intellectual Property Rights and the Emerging Information Infrastructure, *The Digital Dilemma: Intellectual Property in the Information Age* (Washington, D.C.: National Academy Press, 2000).

44. Herbert I. Schiller, *Culture, Inc.* (New York: Oxford University Press, 1989), 32.

45. Thomas Guback, *The U.S. International Film Industry* (Bloomington: Indiana University Press, 1969).

46. Irving Bernstein, *The Economics of Television Film Production and Distribution* (Los Angeles: Screen Actors Guild, 1960), 126; Wilson P. Dizard, *Television: A World View* (Syracuse, N.Y.: Syracuse University Press, 1966), 160–61.

47. For a helpful review of European practice, see Miller, Govil, McMurria, and Maxwell, *Global Hollywood*, 88–99.

48. Dan Schiller and Vincent Mosco, Introduction to *Continental Order? Integrating North America for Cyber-Capitalism*, ed. Vincent Mosco and Dan Schiller (Lanham, Md.: Rowman and Littlefield, 2001), 1–34.

49. Kaarle Nordenstreng, "The MacBride Report: A Milestone in the Great Media Debate," Submission to Quaderno del CAC, Barcelona, on the Twenty-fifth Anniversary of the Approval of the MacBride Report, April 2005, 2 (unpublished manuscript in the author's possession).

50. Daya Thussu quoted in Robert A. Hackett and Yuezhi Zhao, *Democratizing Global Media: One World, Many Struggles* (Lanham, Md.: Rowman and Littlefield, 2005), 8.

51. Dizard, *Television*, 4.

52. Ibid.

53. UNESCO, *Statistical Yearbook 1998* (Paris: UNESCO, 1998), 6.5

54. International Telecommunication Union, *Yearbook of Statistics: Telecommunications Services Chronological Time Series 1989–1998* (Geneva: ITU, 2000).

55. UNESCO, *Statistical Yearbook 1998*, 6.5

56. Feng His-liang, "Wondering about American TV and Newspapers," *Across the Board* 16.3 (March 1979): 3.

57. Oluf Lund Johanson, *World Radio TV Handbook*, vol. 52 (New York: Billboard

Books, 1998), 426, 414, 412, 410; International Telecommunication Union, *Yearbook of Statistics.*

58. United Nations Development Programme, *Human Development Report 2004: Cultural Liberty in Today's Diverse World* (New York: UNDP, 2004), 86.

59. UNESCO, "2001–02—Did You Know? International Trade in Cultural Goods," retrieved 3 January 2005, www.uis.unesco.org.

60. Martin-Barbero quoted in Ana Fiol, "Media and Neoliberalism in Latin America," in *Who Owns The Media? Global Trends and Local Resistances,* ed. Pradip N. Thomas and Zaharom Nain (Penang: Southbound, 2004), 146–47.

61. Herbert I. Schiller, *Mass Communications and American Empire* (New York: Augustus M. Kelley, 1969). For a historical perspective focused on Western Europe, see Victoria de Grazia, *Irresistible Empire: America's Advance through Twentieth-Century Europe* (Cambridge, Mass.: Belknap Press, 2005).

62. John Lawrence Sullivan, "Constructing the Cable Television Market in Latin America: A Structurational Approach to Organizational Knowledge in U.S. Cable Networks" (Ph.D. dissertation, University of Pennsylvania, 2000), 315–17, 80. For other useful accounts, see John Sinclair, *Latin American Television: A Global View* (Oxford: Oxford University Press, 1999), and Rosalind Bresnahan, "The Media and the Neoliberal Transition in Chile," *Latin American Perspectives* 133.6 (November 2003): 39–68.

63. Sullivan, "Constructing the Cable Television Market in Latin America," 87.

64. Ana Fiol, "Media and Neoliberalism in Latin America," 151.

65. McChesney and Schiller, "Political Economy of International Communications," 10.

66. UNESCO, "2001–02—Did You Know?" On games, see Stephen Kline, Nick Dyer-Witheford, and Greig de Peuter, *Digital Play: The Interaction of Technology, Culture, and Marketing* (Montreal: McGill-Queen's University Press, 2003).

67. Mermigas, "More Work Ahead for TW's Parsons," 12–13.

68. Edward S. Herman and Robert W McChesney, *The Global Media: The New Missionaries of Corporate Capitalism* (London: Cassell, 1997), 95.

69. John Sinclair, "Latin American Commercial Television: 'Primitive Capitalism,'" in *A Companion To Television,* ed. Janet Wasko (Oxford: Blackwell Publishing, 2005), 518.

70. Fiol, "Media and Neoliberalism in Latin America," 137.

71. Nicholas Garnham, "Public Service versus the Market," *Screen* 24.1 (January–February 1983): 20.

72. Graham Murdock, "Public Broadcasting and Democratic Culture: Consumers, Citizens, and Communards," in *A Companion to Television,* ed. Janet Wasko (Oxford: Blackwell Publishing, 2005), 178. See also Michael Tracey, *The Decline and Fall of Public Service Broadcasting* (Oxford: Oxford University Press, 1998).

73. Garnham, "Public Service versus the Market," 19.

74. Ibid., 6–27.

75. Ibid., 22.

76. Giuseppe Richeri, "Broadcasting and the Market: The Case of Public Television," in *Toward a Political Economy of Culture: Capitalism and Communication in the Twenty-first Century,* ed. Andrew Calabrese and Colin Sparks (Lanham, Md.: Rowman and Littlefield, 2004), 178–93; Murdock, "Public Broadcasting and Democratic Culture," 190.

77. Tony Barber, "Berlusconi's Critics Take Dim View of TV Privatisation," FT, 29 September 2004, 14; Cinzia Padovani, *A Fatal Attraction: Public Television and Politics in Italy* (Lanham, Md.: Rowman and Littlefield, 2005).

78. Eric Pfanner, "State-Aided Broadcasting Faces Scrutiny across Europe," NYT, 16 February 2004, C1, C7; Charles Goldsmith, "Under Attack, Venerable BBC Is at Crossroads," WSJ, 26 September 2004, B1, B4; Marc Champion, Emily Nelson, and Charles Goldsmith, "To BBC's Rivals, 'Auntie' Is Too Big for Its Britches," WSJ, 28 September 2004, A1, A17; Charles Goldsmith, "At the BBC, a New Austerity," WSJ, 3 December 2004, A17; Alan Cowell, "BBC Says It Will Cut 2,900 Jobs over 3 Years," NYT, 8 December 2004, W1, W7.

79. Bernard Miege, "Capitalism and Communication: A New Era of Society or the Accentuation of Long-Term Tendencies?" in *Toward a Political Economy of Culture: Capitalism and Communication in the Twenty-first Century,* ed. Andrew Calabrese and Colin Sparks (Lanham, Md.: Rowman and Littlefield, 2004), 89.

80. Arthur-Martins Aginam, "Media in 'Globalizing' Africa: What Prospect for Democratic Communication?" in *Democratizing Global Media: One World, Many Struggles,* ed. Robert A. Hackett and Yuezhi Zhao (Lanham, Md.: Rowman and Littlefield, 2005), 127. In Latin America and the Caribbean, by the late 1990s, only sixty-four of about 570 terrestrial broadcast channels were "state-oriented and cultural and educational in scope." Fiol, "Media and Neoliberalism in Latin America," 150.

81. Pradip Thomas, "Contested Futures: Indian Media at the Crossroads," in *Democratizing Global Media: One World, Many Struggles,* ed. Robert A. Hackett and Yuezhi Zhao (Lanham, Md.: Rowman and Littlefield, 2005), 86.

82. Andre Lange, "Transfrontier Television in the European Union: Market Impact," in European Audiovisual Observatory, "Transfrontier Television in the European Union," March 2004, 6, retrieved 3 January 2005, www.obs.coe.int/medium/adv.html .enm.

83. Ibid., 6–7.

84. Amos Thomas, "Flattery or Plagiarism? Television Cloning in India Today," *Media Development* 51.3 (2004): 39.

85. Hugh Miles, *Al Jazeera* (New York: Grove Press, 2005), 329; Nabil H. Dajani, "Television in the Arab East," in *A Companion To Television,* ed. Janet Wasko (Oxford: Blackwell Publishing, 2005), 596.

86. FCC, *In the Matter of Annual Assessment of the Status of Competition in the*

Market for the Delivery of Video Programming, MB Docket No. 04–227, Eleventh Annual Report, 4 February 2005, www.fcc.gov.

87. Dizard, *Television,* 77.

88. Ibid., 277.

89. For a useful overview, see *Communications Satellites: Global Change Agents,* ed. Joseph N. Pelton, Robert J. Oslund, and Peter Marshall (Mahwah, N.J.: Lawrence Erlbaum Associates, 2004).

90. Joseph N. Pelton, "Satellites as Worldwide Change Agents," in *Communications Satellites: Global Change Agents,* ed. Jospeh N. Pelton, Robert J. Oslund, and Peter Marshall (Mahwah, N.J.: Lawrence Erlbaum Associates, 2004), 6.

91. Lange, "Transfrontier Television in the European Union," 11 (table 3).

92. Yuezhi Zhao and Robert A. Hackett, "Media Globalization, Media Democratization: Challenges, Issues, and Paradoxes," in *Democratizing Global Media: One World, Many Struggles,* ed. Robert A. Hackett and Yuezhi Zhao (Lanham, Md.: Rowman and Littlefield, 2005), 21.

93. Naomi Sakr, *Satellite Realms* (London: I. B. Taurus, 2001); Dajani, "Television in the Arab East," 580–601.

94. Miles, *Al-Jazeera,* 348.

95. Ibid., 164–68.

96. Ibid., 275–76.

97. Marc Lynch, *Voices of the New Arab Public: Iraq, al-Jazeera, and Middle East Politics Today* (New York: Columbia University Press, 2006).

98. Miles, *Al-Jazeera,* 278, 324.

99. Jeremy Scahill, "The War on Al Jazeera," *The Nation,* 1 December 2005; retrieved 2 December 2005, www.thenation.com/doc/20051219/scahill.

100. Lynch, *Voices of the New Arab Public,* 5, 6. As Lynch underlines, the growth of Al-Jazeera places pressure on other new Arab satellite stations to permit the airing of some comparable coverage.

101. Miles, *Al-Jazeera,* 426.

102. Ibid., 67; Juan Forero, "And Now, the News in Latin America's View," NYT, 16 May 2005, A8.

103. Nikolas Kozloff, "Chavez Launches Hemispheric, 'Anti-Hegemonic' Media Campaign in Response to Local TV Networks' Anti-Government Bias," Council on Hemispheric Affairs, retrieved 28 April 2005, www.coha.org/NEWPRESSRELEASES/NewPressReleases2005/05.47Telesur %20th one.htm. Thanks to Yuezhi Zhao, Rob Dufy, and Dawn Gable for informing me about Telesur.

104. Ibid., 6, 7; Forero, "And Now, the News in Latin America's View," A8.

105. Quoted in Forero, "And Now, the News in Latin America's View," A8.

106. Andy Webb-Vidal, "Washington to Take Its War of Words with Chavez to Airwaves," FT, 22 July 2005, 7.

107. Sarah Lyall, "BBC to Close 10 Radio Services and Open an Arabic TV Service," NYT, 26 October 2005, A6.

108. Joshua Kucera, "Weaponising the Truth?" *Janes Defence Weekly*, 8 June 2005, 22–25; Adam Brookes, "U.S. Plans to 'Fight the Net' Revealed," BBC News, 27 January 2006; retrieved 1 February 2006, http://news.bbc.co.uk/2/hi/americas/4655196.stm.

109. Jeff Gerth, "Military's Information War Is a Vast and Secret Operation," NYT, 11 December 2005, A1.

110. Miles, *Al-Jazeera*, 277, 323; Edward S. Herman, "'They Kill Reporters, Don't They?'" *Z Magazine* (January 2005): 42–46.

111. Steven R. Weisman, "Under Pressure, Qatar May Sell Arab Network," NYT, 30 January 2005, A1, A17.

112. Miles, *Al-Jazeera*, 425.

113. Eric Pfanner and Doreen Carvajal, "The Selling of Al Jazeera TV to an International Market," NYT, 31 October 2005, C6.

114. James Curran, "The Press as an Agency of Social Control," in *Newspaper History: From the Seventeenth Century to the Present Day*, ed. George Boyce, James Curran, and Pauline Wingate (Beverly Hills, Calif.: Sage Publications, 1978), 51–75.

115. William Wallis, "Al-Jazeera to Launch English Language TV Channel in 2006," FT, 16 May 2005, 6; see also Riz Khan, "Why I'm Joining Al Jazeera," WSJ, 13 June 2005, A12.

116. I draw on Zhao and Hackett, "Media Globalization and Media Democratization," 1; and Miller, Govil, McMurria, and Maxwell, *Global Hollywood*.

117. For a useful exegesis of the *New York Times*'s deception of its readers on the war in Iraq around coverage written by its senior correspondent Judith Miller, see Bill Van Auken, "Burying the Lies on the Iraq War," *Anderson Valley Advertiser* (Boonville, Calif.), 16 November 2005, 8.

118. Diane Mermigas, "Murdoch's Global Health Plan," *Electronic Media*, 10 April 2000, 1, 6; European Commission, Report from the High Level Group on Audiovisual Policy, *The Digital Age: European Audiovisual Policy* (Brussels: European Commission, 1998), 13; Ronald Grover and Tom Lowry, "Rupert's World," *Business Week*, 19 January 2004, 52–60.

119. Daya Kishan Thussu, *International Communication: Continuity and Change* (London: Arnold, 2000), 133–37.

120. Miles, *Al-Jazeera*, 328.

121. U.S. Department of Commerce/International Trade Administration, *U.S. Industry and Trade Outlook 2000* (New York: McGraw-Hill, 2000), 32–34.

122. United Nations Development Programme, *Human Development Report 2004*, 86.

123. Ibid., 97.

124. Elizabeth Guider, "Sony Ups Its Local Payoff," *Variety*, 26 July–1 August 1999, 23.

125. Frank Rose, "Think Globally, Script Locally," *Fortune*, 8 November 1999, 158.

126. Tim Burt and Joshua Chaffin, "Going Where the Growth Is Greater: U.S. Media Groups Reveal Their Global Strategies," FT, 21 June 2005, 15.

127. Anna Wilde Matthews, "A Giant Radio Chain Is Perfecting the Art of Seeming Local," WSJ, 25 February 2002, A1, A10; Sarah Mcbride, "Hit by iPod and Satellite, Radio Tries New Tune: Play More Songs," WSJ, 18 March 2005, A1, A10.

128. Michael J. Wolf, *The Entertainment Economy* (New York: Times Books, 1999), 114.

129. Rose, "Think Globally, Script Locally," 160.

130. Quoted in ibid., 158.

131. Charles Goldsmith, "Moguls Rewrite Script at Cannes as Euro Tanks," WSJ, 19 May 2000, B1.

132. Miller, Govil, McMurria, and Maxwell, *Global Hollywood.*

133. Burt and Chaffin, "Going Where the Growth Is Greater," 15.

134. Schiller and Mosco, Introduction to *Continental Order?* 13–14, 22–23.

135. Cassell Bryan-Low and John Carreyrou, "France's Ubisoft Moves to Rebuff Electronic Arts," WSJ, 6 January 2005, A14; Kline, Dyer-Witherford, and De Peuter, *Digital Play,* 116, 130, 189 (quote).

136. Bruce Orwall, "Colombian Pop Star Taps American Taste in Repackaged Imports," WSJ, 13 February 2001, A1, A6.

137. Burt and Chaffin, "Going Where the Growth Is Greater," 15.

138. These issues are valuably discussed in the context of changes in global children's media in *Pikachu's Global Adventure: The Rise and Fall of Pokemon,* ed. Joseph Tobin (Durham, N.C.: Duke University Press, 2004). See, in particular, Anne Allison, "Cuteness as Japan's Millenial Product," 34–49; and Koichi Iwabuchi, "How 'Japanese' Is Pokemon?" 53–79.

139. Dan Schiller, Enrique Bonus, Meighan Maguire, and Lora Taub, "International Communications and the Struggle for Competitive Advantage in East Asia," in *Changing International Order in North-East Asia and Communications Policies,* ed. Hyeon-Dew Kang (Seoul: NANAM Publishing House, 1992), 73. For the argument that communications was interwoven with U.S. imperial power throughout the post–World War II decades, see Herbert I. Schiller, *Mass Communications and American Empire* (New York: Augustus M. Kelly, 1969).

140. Ellen Meiksins Wood, *Empire of Capital* (London: Verso, 2003).

141. Ibid., 141.

142. Formulation taken partly from ibid., 141.

143. Alan Riding, "A Global Culture War Pits Protectionists against Free Traders," NYT, 5 February 2005, A19.

144. Herbert I. Schiller, *Communication and Cultural Domination* (White Plains, N.Y.: International Arts and Sciences Press, 1976), 24–45.

145. United Nations Development Programme, *Human Development Report 2004,* 96–97.

146. United Nations Educational, Scientific, and Cultural Organization, General Conference, 33d Session, Paris 2005, "Preliminary Report by the Director-General Setting Out the Situation to Be Regulated and the Possible Scope of the Regulating Action Proposed, Accompanied by the Preliminary Draft of a Convention on the Protection of the Diversity of Cultural Contents and Artistic Expressions," Item 8.3 of the Provisional Agenda; retrieved 21 February 2006, www.unesco.org. For a useful contextualizing article, Ted Magder, "Transnational Media, International Trade, and the Idea of Cultural Diversity," *Continuum: Journal of Media and Cultural Studies* 8.3 (September 2004): 380–97.

147. Riding, "Global Culture War Pits Protectionists against Free Traders," A19.

148. Useful information about some of these issues comes from the Internet Governance Project, www.InternetGovernance.org.

149. Christopher Rhoads, "In Threat to Internet's Clout, Some Are Stealing Alternatives," WSJ, 9 January 2006, A1, A7; an overview with links to related sites is found at "Alternative DNS Root," retrieved 19 January 2006, http://en.wikipedia.org/wiki/Alternative DNS-root.

150. "U.S. Will Retain Its Control of Internet Oversight," WSJ, 1 July 2005, B3; Victoria Shannon, "U.S. Seeks to Keep Role on Internet," NYT, 4 July 2005, C6.

151. Associated Press, "Government Asks Internet Agency to Delay '.xxx' Domain Name," WSJ, 17 August 2005, D4.

152. "Controlling Cyberspace" (editorial), *Chicago Tribune*, 14 November 2005, 17.

153. Mark A. Shiffrin and Avi Silberschatz, "Web of the Free," NYT, 23 October 2005, 13.

154. Dennis Byrne, "Mitts Off the Internet, Iran, China, Cuba . . . ," *Chicago Tribune*, 14 November 2005, 17.

155. "e-Meddling" (editorial), WSJ, 17 October 2005, A18.

156. "ITAA Blasts EU for Compromising Internet Governance," axcessnews, 23 October 2005, retrieved 31 October 2005, www.axcessnews.com/modules/wfsection/article.php?/articleid (first quote); Grant Gross, "EU Internet Governance Proposal Raises U.S. Objections," *Infoworld*, 6 October 2005, retrieved 31 October 2005, www.infoworld.com/article/05/10/06/Hneuproposal 1.html (second quote).

157. Gross, "EU Internet Governance Proposal Raises U.S. Objections."

158. Declan McCullagh and Anne Broache, "Bush Challenges EC over Internet Governance," *ZDNET U.K.*, 21 October 2005, retrieved 31 October 2005, http://news.zdnet.co.uk/internet/0,39020369,39232718,00.htm.

159. Frances Williams, "U.S. Retains Control of Global Internet—For Now," FT, 17 November 2005, 4.

160. Saul Hansell, "As Gadgets Get It Together, Media Makers Fall Behind," NYT, 25 January 2006, E1 (first quote), E8 (Leonsis).

161. Ibid., E1.

162. Richard Waters, "Yahoo Feels the Creative Urge," FT, 21 June 2005, 10.

163. Ithiel de Sola Pool and Arthur B. Corte, *The Implications for American Foreign Policy of Low-Cost Non-Voice Communications: A Report to the Department of State* (Cambridge: Center for Policy Alternatives and Center for International Studies, Massachusetts Institute of Technology, 1975), 39 and 45.

164. Martha E. Williams, "The State of Databases Today: 2003," in *Online Databases*, vol. 1 of *Gale Directory of Databases 2003* (New York: Gale Group, 2003) xxiv–xxv.

165. In order, they were: Google, MSN, Microsoft.com, Ebay, and Yahoo. Including Amazon, at number nine, meant that U.S. search engines, portals, and commercial services constitute six of the top ten European destinations. Andy Reinhardt and Robert D. Hof, "Europe Heads for the E-Mall," *Business Week*, 12 July 2004, 51.

166. Stephanie Gruner, "Executives Meet to 'Tame' the Internet, but Critics Fear a Loss of Innovation," WSJ, 13 September 1999, A39.

167. Darin Barney, *Prometheus Wired* (Chicago: University of Chicago Press, 2000), 97–101.

168. Motion Picture Association of America Press Release quoted in Miller, Govil, McMurria, and Maxwell, *Global Hollywood*, 137.

169. Ignacio Romanet, "Final Edition for the Press," *Le Monde Diplomatique*, January 2005, 1; Paul Markillie, "Crowned at Last: A Survey of Consumer Power," *The Economist*, 2 April 2005, 5.

170. National Research Council, Committee on Internet Navigation and the Domain Name System, Computer Science and Telecommunication Board, *Signposts in Cyberspace: The Domain Name System and Internet Navigation*, Prepublication copy (Washington, D.C.: National Academies Press, 2005), ES 10.

171. Carol Pickering, "The World's Local Yokel," *Business 2.0* (May 2000): 188, 193.

172. National Research Council, *Signposts in Cyberspace*, 1–10.

173. Ibid.

174. Thanks to Leigh Estabrook for posing this question.

175. Bob Tedeschi, "E-Commerce Report," NYT, 22 May 2000, C15.

176. Leslie Helm, "The 'World' in World Wide Web Becomes More Visible," *Los Angeles Times*, 22 March 1999, C4.

177. David Block, "Globalization, Transnational Communication, and the Internet," *International Journal on Multicultural Societies* 6.1 (2004): 26, 25.

178. Quoted in Joanne Ingrassia, "Discovery's Web, I-Strategy," *Electronic Media*, 10 April 2000, 50.

179. Malini Guha, "Chinese-Language Pop Music Finds Online Outlets," FT, 7 February 2005, 2.

180. Darin Barney, *The Network Society* (Maldin, Mass.: Polity, 2004), 173.

181. Yuezhi Zhao and Dan Schiller, "Dances with Wolves? China's Integration with Digital Capitalism" *Info* 3.2 (April 2001): 137–51.

182. Kevin Robins and Frank Webster, "Cybernetic Capitalism: Information, Technology, Everyday Life," in *The Political Economy of Information*, ed. Vincent Mosco and Janet Wasko (Madison: University of Wisconsin Press, 1988), 44–75.

183. For an especially insightful study, see Ellen Seiter, *The Internet Playground* (New York: Peter Lang, 2005).

184. James Cortada, *The Digital Hand* (New York: Oxford University Press, 2004); Dan Schiller, *Telematics and Government* (Norwood, N.J.: Ablex, 1982). For a recent journalistic account, see Yochi J. Dreazen, Grep Ip, and Nicholas Kalish, "Why the Sudden Rise in the Urge to Merge and Form Oligopolies?" WSJ, 25 February 2002, A1, A10.

Chapter 7: Parasites of the Quotidian

1. Raymond Williams, *Problems in Materialism and Culture* (1961; reprint, London: Verso, 1983), 183.

2. United Nations Development Programme, *United Nations Human Development Report 1998* (New York: Oxford University Press, 1998), 63; see also Stuart Elliott, "Advertising," NYT, 5 December 2000, C12.

3. Stuart Elliott, "No More Same-Old," NYT, 23 May 2005, C1, C8.

4. Stuart Elliott, "Advertising," NYT, 23 June 2004, C5; Robyn Greenspan, "Optimism Drives Ad Forecast Revisions," *ClickZStats*, 24 June 2004, retrieved 3 December 2004, www.clickz.com/stats/sectors/advertising/article.php/3373311; Stuart Elliott, "Advertising," NYT, 18 April 2005, C10.

5. Bob Davis, "Lagging behind the Wealthy, Many Use Debt to Catch Up," WSJ, 17 May 2005, A1, A10.

6. Judann Pollack and Mercedes M. Cardona, "Kraft to Boost Marketing; Plans New-Product Blitz," *Advertising Age*, 5 July 1999, 10.

7. Mark Maremont, "Gillette's Venus Razor for Women to Be Born Amid a Big Ad Drive," WSJ, 3 November 2000, B5.

8. Kerry Capell, "Novartis' Marketing Doctor," *Business Week*, 5 March 2001, 56; Barry Meier and Stephanie Saul, "Marketing of Vioxx: How Merck Played Game of Catch-Up," NYT, 11 February 2005, A1, C2; Marcia Angell, "The Truth about the Drug Companies," *New York Review of Books*, 15 July 2004, 52–58.

9. Brian Steinberg, "P&G Brushes Up Iconic Image of 'Crest Kid' in New Campaign," WSJ, 29 March 2005, B9.

10. "In the Giants' Long Shadow," *Advertising Age*, 14 February 2000, 30; Erin White, "WPP Deal Puts Pressure on Havas," WSJ, 14 September 2004, B14.

11. Anthony Bianco, "The Vanishing Mass Market," *Business Week*, 12 July 2004, 60–72.

12. Stuart Elliott, "How Agencies Read Signs of a Slowdown," NYT, 26 February 2001, C1, C16.

13. Richard Tomkins, "Interpublic Hired to Define Coke Globally," FT, 4 December 2000, 20.

14. Allen Rosenshine, "Evolving Agencies' Mission," *Advertising Age*, 8 November 1999, 8.

15. Dean Foust, "Gone Flat," *Business Week*, 20 December 2004, 76–82.

16. Elliott, "No More Same-Old," C1, C8.

17. Ibid.

18. Scott Hensley, "Some Drug Makers Are Starting to Curtail TV Ad Spending," WSJ, 16 May 2005, B1, B8; Joe Flint and Brian Steinberg, "Ad Icon P&G Cuts Commitment to TV Commercials," WSJ, 13 June 2005, A1, A6.

19. Herbert I. Schiller, *Who Knows: Information in the Age of the Fortune 500* (Norwood, N.J.: Ablex, 1981).

20. Constance L. Hays, "Broadcasting to a Captive Audience," NYT, 21 February 2005, C1, C6.

21. Ethan Smith, "Befitting Its Name, Song Airlines Becomes Music Promoter," WSJ, 26 May 2005, B1, B2.

22. Kelefa Sanneh, "Alanis Revisits Her Hits with a Singalong," NYT, 17 June 2005, B3.

23. Elliott, "No More Same-Old," C1, C8.

24. Ibid., C1, C8.

25. H. H. Wilson, *Pressure Group: The Campaign for Commercial Television* (London: Secker and Warburg, 1961).

26. Armand Mattelart, *Advertising International: The Privatisation of Public Space* (London: Comedia, 1991).

27. Betsy McKay, "Coke's 'Think Local' Strategy Has Yet to Prove Itself," WSJ, 1 March 2001, B6.

28. Leslie Cauley, "Discovery Is Tailoring 'Cleopatra' as a One-Stop Global Media Buy," WSJ, 10 March 1999, B17.

29. Bruce Orwall, "MGM Sets Accord with MTV Network for New Bond Film," WSJ, 29 September 1999, B8.

30. Geoffrey A. Fowler, "McDonald's Asian Marketing Takes On a Regional Approach," WSJ, 26 January 2005, B3.

31. David A. Aaker and Erich Joachimsthaler, "The Lure of Global Branding," *Harvard Business Review* 77.61 (November–December 1999): 137–44.

32. Fowler, "McDonald's Asian Marketing Takes On a Regional Approach," B3.

33. United Nations Development Programme, *Human Development Report 1998*, 63.

34. Michael Flagg, "Vietnam Opens Industry to Foreigners," WSJ, 28 August 2000, B8.

35. "Ad Spending Rose in Japan Last Year, Snapping Slump," WSJ, 18 February 2005, B3; Geoffrey A. Fowler, "Chinese Game Show Offers a Big Prize: A 15–Second Ad Slot," WSJ, 30 November 2004, A1, A12; Geoffrey A. Fowler, "Dentsu Sees Flat Ads Sales in China," WSJ, 15 June 2005, B5; Geoffrey A. Fowler and Juying Qiu, "China Goes from Torrid to Just Hot," WSJ, 17 February 2006, B3.

36. Wilkofsky Gruen Associates, "TV Advertising in Europe," *Electronic Media*, 27 November 2000, 12.

37. Mattelart, *Advertising International*, ix.

38. Dan Schiller, *Digital Capitalism: Networking the Global Market System* (Cambridge: Massachusetts Institute of Technology Press, 1999), 89–142.

39. Mylene Mangalindan, "After Wave of Disappointments, the Web Lures Back Advertisers," WSJ, 25 February 2004, A1, A6; Kevin J. Delaney, "Internet Ads Click with Firms; Some Shift Budgets," WSJ, 3 May 2005, B8.

40. Joseph Pereira, "Junk-Food Games," WSJ, 3 May 2004, B1, B4.

41. Tom Zeller Jr., "Law to Bar Junk E-Mail Encourages Flood Instead," NYT, 1 February 2005, A1, C8.

42. Suzanne Vranica, "Postal Service Touts Direct Mail," WSJ, 15 February 2005, B4.

43. Riva Richmond, "Blogs Keep Internet Customers Coming Back," WSJ, 1 March 2005, B9; Brian Steinberg, "Corporate Marketers Try Out Blogs," WSJ, 3 May 2005, B8 (quote).

44. Brian Steinberg, "Marketing Folks' New Medium May Be Your PC's Hard Drive," WSJ, 2 May 2005, B8; Ronald Grover, "Mad Ave. Is Starry-Eyed over Net Video," *Business Week*, 23 May 2005, 36–39.

45. Brian Steinberg and Suzanne Vranica, "The Ad World's Message for 2005: Stealth," WSJ, 30 December 2004, B1, B3; Brock Read, "College Students Are Bombarded by Cellphone 'Spim,' Study Finds," *Chronicle of Higher Education*, 8 April 2005, A30.

46. Cynthis H. Cho, "For More Advertisers, the Medium Is the Text Message," WSJ, 2 August 2004, B1, B4.

47. Abigail Zuger, "Fever Pitch: Getting Doctors to Prescribe Is Big Business," NYT, 11 January 1999, A1, A13. For a more comprehensive assessment, see U.S. General Accounting Office, "Prescription Drugs FDA Oversight of Direct-to-Consumer Advertising Has Limitations," GAO-03–177, October 2002, www.gao.gov/new.items/d03177.pdf.

48. For an important study, see Inger Stole, *Advertising on Trial: Consumer Activism and Corporate Public Relations in the 1930s* (Urbana: University of Illinois Press, 2006).

49. Sarah Ellison, "Divided, Companies Fight for Right to Plug Kids' Food," WSJ, 26 January 2005, B1, B2.

50. Schiller, *Who Knows;* and Herbert I. Schiller, *Culture, Inc.* (New York: Oxford University Press, 1989).

51. "It's Not Just Nike's Free Speech at Issue" (editorial), *Champaign-Urbana (Ill.) News-Gazette*, 20 April 2003.

52. Stephen Kiehl, "The Art of Brand-Name Dropping," *Los Angeles Times*, 25 August 2004, E5.

53. Joe Flint, "CBS Eyes Windfall from Record Rates for 'Survivor' Ads," WSJ, 12 January 2001, B4.

54. Shelly Branch, "Product Plugs—'M'm M'm Good'?" WSJ, 14 November 2000, B1.

55. Brooks Barnes, "A Good Soap Script Includes Love, Tears, and Frosted Flakes," WSJ, 17 January 2005, A1, A8.

56. Suzanne Vranica, "Miller Calls the Shots in FX Drama Deal," WSJ, 13 October 2004, B1, B5.

57. Quoted in Flint and Steinberg, "Ad Icon P&G Cuts Commitment to TV Commercials," A6.

58. Robert Guy Matthews, "London Stage Hosts U.S. Marketers," WSJ, 18 February 2005, B3.

59. Geoffrey A. Fowler, "In Asia, It's Nearly Impossible to Tell a Song from an Ad," WSJ, 31 May 2005, A1, A10.

60. Brian Steinberg, "'Star Wars' Tie-Ins May Lose Force," WSJ, 10 May 2005, B3.

61. Michael Schroeder, "Some Professors Take Payments to Express Views," WSJ, 10 December 2004, B1, B3.

62. James Bandler, "How Companies Pay TV Experts for On-Air Product Mentions," WSJ, 19 April 2005, A1, A12.

63. Maureen Magee, "More Schools Ask Students: 'Want Fries with That?'" San Diego Union-Tribune, 17 February 2000, A1, A19.

64. June Kronholz, "Economic Time Bomb: U.S. Teens Are Among Worst at Math," WSJ, 7 December 2004, B1, B6; Constance L. Hays, "Math Book Salted with Brand Names Raises New Alarm," NYT, 21 March 1999, 1.

65. Schiller, Culture, Inc.; Chin-tao Wu, Privatising Culture: Corporate Art Intervention since the 1980s (London: Verso, 2002); Michelle Falkenstein, "And Now a Word about the Sponsors," Art News 99.5 (May 2000): 218–22.

66. Alessandra Stanley, "Modern Marketing Blooms in Medieval Vatican Library," NYT, 8 January 2001, A1, A6.

67. Edmund Sanders, "Postal Service Putting Its Stamp on Ads to Pay Bills," Los Angeles Times, 24 February 2001, A1; Jessica Mintz, "Your Face on a Stamp Again? Custom Photo Postage Is Back," WSJ, 27 April 2005, B1, B2.

68. Robert W. McChesney, "On Media and the Election," The Public I 4.10 (December 2004), 1; Katharine Q. Seelye, "How to Sell a Candidate to a Porsche-Driving, Leno-Loving Nascar Fan," NYT, 6 December 2004, A16.

69. Christopher Cooper, "Marketing Nirvana Is to Be a President's Preferred Brand," WSJ, 18 April 2005, B1, B6.

70. A. Craig Copetas, "Soccer Teams Study Stadium Branding," WSJ, 24 April 2000, A11.

71. Edwin McDowell, "The Parade of Corporate Sponsors," NYT, 16 July 1999, C1, C17.

72. Stefan Fatsis and Joe Flint, "CBS, Fox, and DirecTV Signs NFL Deals Totaling $11.5 Billion," WSJ, 9 November 2004, B3.

73. Marilyn Chase, "Do Sponsors Sway Health Web Sites?" WSJ, 8 February 2000, B7.

74. William L. Bulkeley, "New England Journal Editor Blasts Some Drug Industry/Academic Links," WSJ, 18 May 2000, B18.

75. Paul Raeburn, "The Corruption of TV Health News," Business Week, 28 February 2000, 66–68.

76. Claudia Eller and Sallie Hofmeister, "USA Network Cancels Film after Advertiser Protests," Los Angeles Times, 6 December 2000, A1, A32.

77. Meier and Saul, "Marketing of Vioxx," A1, C2.

78. Anna Wilde Matthews, "Worrisome Ailment in Medicine: Misleading Journal Articles," WSJ, 10 May 2005, A1, A9.

79. Kevin Helliker, "Yet Another Reason to Go to the Gym: How Exercise Can Help Fight Depression," WSJ, 10 May 2005, D1.

80. David Hatch, "Local News Execs Feeling Ad Pressure," Electronic Media, 1 December 2000, 37.

81. Brian Steinberg and Joseph T. Hallinan, "GM Has Little to Gain in Paper Fight," WSJ, 11 April 2005, B5.

82. Katherine Q. Seelye, "Print Media Work to Convince Advertisers They Still Matter," NYT, 2 May 2005, C4.

83. Elizabeth Weinstein, "Graffiti Cleans Up at Retail," WSJ, 12 November 2004, B1, B3.

84. Eduardo Kaplan, "Che Guevara Rebranded for Retail Sparks a Host of Contradictions," WSJ, 23 September 2004, A12.

85. Amy S. Rosenberg, "Baby on Block: Imagine Your Logo on Her Newborn," Philadelphia Inquirer, 28 May 2005. Thanks to Edward Herman for this reference.

86. Geoffrey A. Fowler, "Chinese Youth League Turns to a New Path: Madison Avenue," WSJ, 25 January 2005, A1, A13.

87. Karby Leggett, "Pay-as-You-Go Kibbutzim," WSJ, 26 May 2005, B1, B2.

88. Gary Silverman, "FAA Kicks Billboard Idea into Orbit," FT, 21 May 2005, 8.

89. Jonathan Kaufman, "Marketing in the Future Will Be Everywhere—Including Your Head," WSJ, 1 January 2000, R26.

90. Stephen Kline, Nick Dyer-Witheford, and Greig de Peuter, Digital Play: The Interaction of Technology, Culture, and Marketing (Montreal: McGill-Queen's University Press, 2003), 236, 221.

91. A useful overview of television ratings may be found in Eileen R. Meehan, "Watching Television: A Political Economic Approach," in A Companion to Television, ed. Janet Wasko (Oxford: Blackwell Publishing, 2005), 238–55. For the wider context of corporate gathering of personal information, see Oscar Gandy, The Panoptic Sort (Boulder, Colo.: Westview, 1996).

92. Victoria de Grazia, Irresistible Empire: America's Advance through Twentieth-Century Europe (Cambridge, Mass.: Belknap Press, 2005), 101.

93. Toby Miller, Nitin Govil, John McMurria, and Richard Maxwell, Global Hollywood (London: BFI, 2001), 206.

94. Stuart Elliott, "Hoping to Quell Concerns about the Accuracy of Its TV Ratings Data, Nielsen Presents a Research Plan," NYT, 22 February 2005, C6.

95. Elizabeth Weise, "Companies Learn Value of Grass Roots," USA Today, 26 May 1999, 4D.

96. Gillian Tett, "Lost Tribes of Acme Accounting," FT, 21 May 2005, W1, W2.

97. Joseph Pereira, "Spying on the Sales Floor," WSJ, 21 December 2004, B1, B4.

98. Gary Silverman, "Back to the Future: Advertisers Get out of the Living Room and on to the Street," FT, 25 October 2004, 11.

99. Gary Silverman, "Why the Boardroom Believes in Reality TV," FT, 1 March 2005, 11.

100. Jennifer Lee, "Caught on Videotape, and Then Simply Caught," NYT, 22 May 2005, 29.

101. Jason Anders, "Web-Filter Data from Schools Put Up for Sale," WSJ, 26 January 2001, B1, B4.

102. Kathryn Kranhold, "Database Is Key to Agency Stake in Medical Sites," WSJ, 24 August 2000, B1, B16.

103. Kathryn Kranhold and Michael Moss, "Keep Away from My Cookies, More Marketers Say," WSJ, 20 March 2000, B1, B6.

104. Nick Wingfield and Glenn R. Simpson, "With So Much Subscriber Data, AOL Walks a Cautious Line on Privacy," WSJ, 15 March 2000, B1.

105. Edmund Sanders, "Media Giant Serving Two Masters," Los Angeles Times, 14 February 2001, C1, C5.

106. "Coke Plans to Donate 50 Years of TV Spots to Library of Congress," WSJ, 29 November 2000, B11.

Chapter 8: Mobilized

1. James E. Katz and Mark Aakhus, Perpetual Contact: Mobile Communication, Private Talk, Public Performance (New York: Cambridge University Press, 2002).

2. David Pringle, "Motorola Trails Rivals' Growth in Cellphones," WSJ, 4 February 2004, B5; Phred Dvorak, "DoCoMo's I-Mode Growth Slows at Home," WSJ, 4 March 2004, B4; Jesse Drucker and Karen Lundegaard, "As Industry Pushes Headsets in Cars, U.S. Agency Sees Danger," WSJ, 19 July 2004, A1, A7; Kevin J. Delaney, "Text Messaging May Be Peaking in Europe," WSJ, 16 October 2003, B4, B6; "China's Cyber Censors" (editorial), WSJ, 6 July 2004, A22.

3. Greg Ip and Mark Whitehouse, "Huge Flood of Capital to Invest Spurs World-Wide Risk Taking," WSJ, 3 November 2005, A1, A6.

4. Quoted in Erika Brown, "Coming Soon to a Tiny Screen Near You," Forbes, 23 May 2005, 67.

5. Matt Richtel, "$50 Million Is Raised for Venture in Wireless," NYT, 13 June 2005, C10.

6. Eleanor Randolph, "The Cell Tower Blight: Text-Message Caller, ASAP," NYT, 26 February 2005, A26.

7. Brown, "Coming Soon to a Tiny Screen Near You," 67.

8. Ibid., 67, 68.

9. David Pringle and Don Clark, "Nokia, Intel Plan to Collaborate on Wireless Technology WiMAX," WSJ, 10 June 2005, B3.

10. Christopher Rhoads and Nick Wingfield, "Apple's iPod Faces Challenge from Cellphones," WSJ, 11 April 2005, B1, B4.

11. David Pringle and Charles Goldsmith, "Now Hear This: Cellphone Remixes," WSJ, 18 March 2004, B5.

12. Matt Richtel, "Makers of Cellphone Video Games Suddenly Find Great Expectations," NYT, 16 May 2005, C12.

13. Stephanie N. Mehta, "Phone Companies See a Wireless Future," WSJ, 15 November 1999, A6.

14. "Why You *Still* Can't Hear Me Now," WSJ, 25 May 2005, D1, D6.

15. David Pringle, Jesse Drucker, and Evan Ramstad, "Cellphone Makers Pay a Heavy Toll for Missing Fads," WSJ, 30 October 2003, A1, A10.

16. Jessica Tan, "Motorola Targets Top Spot in China," WSJ, 15 June 2005, B3.

17. Cell phones contain toxins including lead, cadmium, mercury, and materials that can degrade into arsenic. Lisa Guernsey, "Phones in the Drawer or in the Trash, or to a Good Cause," NYT, 28 February 2002, D7; Jesse Drucker, "Old Cellphones Pile Up by the Millions," WSJ, 23 September 2004, B1, B6.

18. Ginny Parker, "DoCoMo Feels Heat of Rival," WSJ, 17 May 2004, B4.

19. Dvorak, "DoCoMo's I-Mode Growth Slows at Home," B4; Ken Belson, "Fast Phones in Japan, but No Pot of Gold," NYT, 6 June 2005, C8.

20. Belson, "Fast Phones in Japan, but No Pot of Gold," C8; David Pringle, "Slower Growth Hits Cellphone Services Overseas," WSJ, 23 May 2005, A1, A6.

21. Pringle, "Slower Growth Hits Cellphone Services Overseas," A1.

22. David Pringle, "Vodafone's Services Take Hold," WSJ, 19 November 2003, B8.

23. David Pringle, "High-Tech Cellphones Catch On in Europe as Models Get Lighter," WSJ, 1 June 2004, B4; Paul Taylor, "Dialing for Dollars with Data," FT, 2 July 2004, 17.

24. Cassell Bryan-Low and David Pringle, "Sex Cells," WSJ, 12 May 2005, B1, B2.

25. David Kessmodel, "Nevada May Allow Hand-Held Gaming Devices," WSJ, 24 May 2005, B1, B2.

26. Pringle, "Slower Growth Hits Cellphone Services Overseas," A6.

27. David Pringle, "Vodafone Looks to Emerging Markets," WSJ, 10 March 2005, A3.

28. Nic Fildes, "Mobile-Phone Service Benefits Africa," WSJ, 16 February 2005, B4.

29. Belson, "Fast Phones in Japan, but No Pot of Gold," C8.

30. Ryan J. Foley, "Consumer Complaints Soared in 2002," WSJ, 25 November 2003, D2.

31. Jesse Drucker, "Cellphone Bills See New Round of Hidden Fees," WSJ, 1 May 2003, D1, D3; "Three Steps to Better Cellular," Consumer Reports 68.2 (February 2003): 15–16.

32. Simon Romero, "Cellphone Service Hurt by Success," NYT, 18 November 2002, A1, A17.

33. U.S. General Accounting Office, Report to the Honorable Anthony D. Weiner, House of Representatives, "Telecommunications: FCC Should Include Call Quality in Its Annual Report on Competition in Mobile Phone Services," April 2003, retrieved 6 February 2006, www.gao.gov/cg-bin/getrpt?GAO-03–501.

34. "Why You Still Can't Hear Me Now," D1, D6.

35. Romero, "Cellphone Service Hurt by Success," A1, A17.

36. Dennis K. Berman and Jesse Drucker, "Static, Problems Are Hampering Wireless Growth," WSJ, 24 November 2003, B1, B4.

37. Jesse Drucker, "Spotty Cellphone Service Frustrates Customers," WSJ, 18 August 2003, B4.

38. "Why You Still Can't Hear Me Now," D1, D6. One of the early contractors for the Arpanet—the packet-switched system that was a forerunner of today's Internet—explicitly sought to provide the reliability of "the phone company." Janet Abbate, Inventing the Internet (Massachusetts Institute of Technology Press, 1999), 65.

39. Berman and Drucker, "Static, Problems Are Hampering Wireless Growth," B1, B4.

40. Drucker, "Spotty Cellphone Service Frustrates Customers," B4; Andrew Ross Sorkin and Matt Richtel, "Cellphone Failures Cause Many to Question Systems," NYT, 16 August 2003, B7; Yochi J. Dreazen, "Panicked Phone Traffic Jams Lines in Northeast," WSJ, 12 September 2001, A3.

41. Quoted in Berman and Drucker, "Static, Problems Are Hampering Wireless Growth," B1, B4.

42. Romero, "Cellphone Service Hurt by Success," A1, A17.

43. U.S. General Accounting Office, Report to the Honorable Anthony D. Weiner, House of Representatives, "Telecommunications."

44. Peter Grant, "Phone System's Weak Link," WSJ, 17 September 2004, B1, B2; Jesse Drucker and Amy Schatz, "Phone Outages Grow More Severe," WSJ, 10 November 2005, B10.

45. Anne Marie Squeo, "Cellphone Hangup: When You Dial 911, Can Help Find You?" WSJ, 12 May 2005, A1, A10 (quote); see also Anne Marie Squeo, "Tests Show Many Cellphone Calls to 911 Go Unlocated," WSJ, 19 May 2005, B1, B6.

46. Christopher Conkey, "Do-Not-Call Lists under Fire," WSJ, 28 September 2005, D1, D3.

47. Quoted in Matt Richtel, "For Now, Unwired Means Unlisted; That May Change,"

NYT, 30 August 2004, C1, C6; Jesse Drucker, "Phone Directory of Cell Numbers Creates Static," WSJ, 14 January 2005, B1, B8.

48. Raymond Williams, *Television: Technology and Cultural Form* (London: Fontana, 1974), 26.

49. Carolyn Marvin, *When Old Technologies Were New: Thinking about Communications in the Late Nineteenth Century* (New York: Oxford University Press, 1988).

50. Raymond Flandez, "Cell Use Leads Schools to Alter Phone Options," WSJ, 7 August 2003, B1; Matt Richtel, "School Cellphone Bans Topple (You Can't Suspend Everyone)," NYT, 29 September 2004, A1, A16; Fox Butterfield, "Inmates Use Cellphones to Maintain a Foot on the Outside," NYT, 21 June 2004, A1, A18; Matt Richtel, "For Liars and Loafers, Cellphones Offer an Alibi," NYT, 26 June 2004: A1, A4.

51. Josephine Ma, "Migrant Work Still Fraught with Risk," *South China Morning Post*, 19 June 2004.

52. Organization for Economic Cooperation and Development, *OECD Employment Outlook 2004*, chapter 1, chart 1.3, retrieved 17 July 2004, www.oecd.org. Trends in hours per capita have diverged since 1970.

53. Employed adult women spent about an hour more per day than employed adult men doing household activities and caring for household members. U.S. Department of Labor, Bureau of Labor Statistics, "Time-Use Survey—First Results Announced by BLS," 14 September 2004, retrieved September 2004, www.bls.gov./tus/.

54. U.S. Department of Transportation, Bureau of Transportation Statistics, "America on the Go . . . Findings from the National Household Travel Survey: U.S. Business Travel," retrieved 6 February 2006, http://bts.gov/publications/america on the go/us business travel/pdf/entire.pdf.

55. U.S. Department of Transportation, Bureau of Transportation Statistics, "National Household Travel Survey, 2001" retrieved 6 February 2006, www.bts.gov/publications/highlights of the 2001 national household travel survey/pdf/entire.pdf.

56. Ibid.

57. Ibid.

58. "Stuck In Traffic," NYT, 10 May 2005, A16.

59. Drucker and Lundegaard, "As Industry Pushes Headsets in Cars, U.S. Agency Sees Danger," A1, A7; Karen Lundegaard and Jesse Drucker, "Cellphones Are Found to Pose Riskiest Distractions for Drivers," WSJ, 15 June 2005, D4.

60. "Cellphone Gabbing behind the Wheel Mounted Last Year," WSJ, 2 March 2005, D3.

61. "Cost of Cellphone-Tied Crashes Equals Benefit of Calls from Cars," WSJ, 2 December 2002, A8; Harvard Center for Risk Analysis, "Updated Study Shows Higher Risk of Fatality from Cell Phones While Driving," retrieved 6 February 2006, http://www.hcra.harvard.edu/cellphones.html.

62. Lundegaard and Drucker, "Cellphones Are Found to Pose Riskiest Distractions for Drivers"; Robert W. Hahn and James E. Prieger, "The Impact of Driver Cell Phone Use on Accidents," American Enterprise Institute/Brookings Joint Center for Regula-

tory Studies, Working Paper 04–14, July 2004, retrieved 6 February 2006, http://www
.aei-brookings.org/publications/abstract.php?pid=806; Karen Lundegaard and Jesse
Drucker, "Cellphone Use Raises Car-Crash Risk," WSJ, 12 July 2005, D5.

63. "Are Cell Phones Safe?" 24.

64. "Global Insights from Internet Untethered Participants" (advertisement), NYT,
21 August 2000, 16; Ryan J. Foley, "Back Door Could Open to Cellphone Telemarketing," WSJ, 18 November 2003, D4.

65. Sarmad Ali, "Cellphone Firms Accused of Billing for Unwanted Ads," WSJ, 21
July 2005, D5.

66. Gren Manuel, "Dial-a-Coke Is Slaking Thirsts in Australia," WSJ, 20 October
2003, R3.

66. Gren Manuel, "Dialing for Dollars," WSJ, 20 October 2003, R3.

67. Amy Harmon, "Lost? Hiding? Your Cellphone Is Keeping Tabs," NYT, 21 December 2003, 1, 28.

68. Lee, "Space, Mobile Tracking, and Workers," 11.

69. Matt Richtel, "Live Tracking of Mobile Phones Prompts Court Fights on Privacy," NYT, 10 December 2005, A1, B13.

70. Kwang-Suk Lee, "Space, Mobile Tracking, and Workers: The Case of Samsung,"
(Unpublished research paper in the author's possession, 2005), 14.

71. Quoted in Barnaby J. Feder, "What's in the Box? Radio Tags Know That, and
More," NYT, 27 September 2004, C4; Barnaby J. Feder, "IBM Expands Effforts to
Promote Radio Tags to Track Goods," NYT, 14 June 2005, C9.

Chapter 9: Open Questions about China, Information, and the World Economy

1. "Losing Its Balance" (editorial), *The Economist*, 20 March 2004, 12.

2. Matt Pottinger, "Xinhua Seeks Asia News Unit Owned by French Press Agency,"
WSJ, 11 December 2002, B5.

3. Susan V. Lawrence, "Broadcast News, Chinese Style," *Far Eastern Economic
Review*, 2 May 2002, 30–31; Ting Shi, "AOL Time Warner/News Corp. Deal: Win-Win
or Not," retrieved January 2003, http://journalism.Berkeley.edu/projects/asiaprojects/
shi.html.

4. Angela Mackay, "Tom.com Buys Stake in AOL China TV Unit," FT, 3 July 2003,
27.

5. M. Dickie and A. Harney, "China Ready to Relax Regulations on Foreign Investment in TV Production," FT, 15 November 2004, 1.

6. Evan Ramstad and Kevin J. Delaney, "Thomson to Create Venture with China's
TCL to Make TVs," WSJ, 3 November 2003, B4.

7. Winston Yau, "TV Makers Win Tariff Fight: Trade Breakthrough Clears Way
for Cheap Chinese Exports to Flood European Markets," *Business Post*, 5 September
2002, 1.

8. Ramstad and Delaney, "Thomson to Create Venture with China's TCL to Make TVs," B4; Justine Lau and Doug Cameron, "TCL May Float 25% of Its TV joint Venture," FT, 4 November 2003, 17.

9. Evan Ramstad, "East Meets West in TV Sets," WSJ, 26 November 2004, A7.

10. Gabriel Kahn, "Yo, Yao: What's Up with Chinese Ads in Texas?" WSJ, 7 February 2003, B1, B4.

11. Peter Wonacott, "Yao-Mania: Hoop Star's China Visit Evokes Beatles, 1964," WSJ, 15 October 2004, B1, B7.

12. G. A. Fowler, "Chinese Game Show Offers a Big Prize: A 15–Second Ad Slot," WSJ, 30 November 2004, A1, A12; Brook Larmer, "The Center of the World," Foreign Policy (September–October 2005): 66–74.

13. "HSBC and Partner Offer Credit Card in China," NYT, 5 November 2004, W1; Andrew Browne, "Wooing China's Credit-Card Users," WSJ, 30 October 2005, C1, C3.

14. Andrew Batson, "China's Cellphone Output Soars," WSJ, 31 July 2003, B3; Asher Bolande, "Handsets from China Driving Down Prices," WSJ, 30 January 2003, B6; Jesse Drucker, David Pringle, and Evan Ramstad, "Pricing Pressure Squeezes Cellphone Makers World-Wide," WSJ, 15 January 2004, B1, B6; Evan Ramstad, "China's Makers of Cellphones Thrive at Home," WSJ, 21 August 2003, B1, B4; Evan Ramstad, "China's Cellphone Market Has Even More Room to Grow," WSJ, 20 October 2003, B7; Evan Ramsted, "China Has Cellphone Hangover," WSJ, 2 September 2003, B10.

15. Bolande, "Handsets from China Driving Down Prices," B6.

16. Evan Ramstad, "In Tech, China Is Setting the Standard," WSJ, 10 September, A22; Kathy Chen, "China Sets Own Wireless Encryption Standard," WSJ, 3 December 2003, B4; Kathy Chen, "China Sets Limits on Wireless Sales," WSJ, 10 December 2003, B6; Neil King Jr., "China Urged to Drop Tech Rule," WSJ, 4 March 2004, B4; "China Rolls Out DVD Alternative Called the EVD," WSJ, 19 November 2003, B8.

17. Richard Waters, "People Power to Make China Online Capital," FT, 29 November 2004, 7; Rebecca Buckman, "China to Open Telecom Sector but Few Foreign Players Rush In," WSJ, 9 December 2004, B5.

18. "Slim Pickings," The Economist, 20 March 2004, 64; J. L. Schenker, "Asians Rush to Buy Assets of Ailing Giants: U.S. Telecom Pain Is World's Gain," International Herald-Tribune, 25 August 2004, retrieved 6 February 2006, http://www.iht.com/articles/535472.htm.

19. Henry Sender, "Asia Global Crossing Bid Signals Chinese Arrival as Merger Players," WSJ, 19 November 2002, C5; Asher Bolande and Dennis Berman, "China Netcom Group to Buy Network," WSJ, 18 November 2002, A3; Joe Leahy, "Saved Asia Netcom Aims for Break-Even," FT, 12 March 2003, 18; Schenker, "Asians Rush to Buy Assets of Ailing Giants."

20. Mei Fong, "China Netcom Sets Its Sights Abroad," WSJ, 26 February 2004, B5; Han Dong, "From China Unicom to China Netcom: A History of Corporate Transforming in China's Telecommunications Industry" (Unpublished manuscript

in the author's possession, December 2004), 22; Leahy, "Saved Asia Netcom Aims for Break-Even," 18; F. Guerrera and P. Taylor, "Netcom Price Below Expectations," *FT*, 27 October 2004, 19. All three other major Chinese carriers had already sought such listings, including Netcom's much larger rival, China Telecom. Bolande and Berman, "China Netcom Group to Buy Network," A3.

21. Almour Latour and Matt Pottinger, "Can Li Ka-Shing Make New Phones Ring Up Big Profit?" *WSJ*, 20 November 2002, A1, A13; Evan Ramstad, Nisha Gopalan, and C. K. Sing, "Hutchison Whampoa Cellular Service Takes Off," *WSJ*, 19 March 2004, B4; Rebecca Buckman and David Pringle, "3G Phones Don't Impress Brits," *WSJ*, 16 December 2004, B1, B2.

22. Jason Dean, "Long a Low-Tech Power, China Sets Its Sights on Chip Making," *WSJ*, 17 February 2004, A1, A16.

23. Jason Dean, "Entrepreneurs Bet on Chip Designing in China," *WSJ*, 2 December 2004, B4, B6.

24. Jason Dean, "China Considers Venture Fund for Chip Industry," *WSJ*, 26 November 2004, B2.

25. Allan R. Gold, Glenn Leibowitz, and Anthony Perkins, "A Computer Legend in the Making," *McKinsey Quarterly* 3 (March 2003): 1, 7.

26. Charles Hutzler, "In China, Turf Battle Rages," *WSJ*, 29 June 2004, A12.

27. Keith Bradsher, "Chinese Computer Maker Plans a Push Overseas," *NYT*, 22 February 2003, B1, B3; Rebecca Buckman, "Computer Giant in China Sets Sights on U.S.," *WSJ*, 18 June 2003, B1, B4; Evan Ramstad and Gary McWilliams, "For Dell, Success in China Tells Tale of Maturing Market," *WSJ*, 5 July 2005, A1, A8.

28. David Barboza, "An Unknown Giant Flexes Its Muscles," *NYT*, 4 December 2004, B1, B3; G. Rivlin and John Markoff, "Weighing IBM's Possible Absence in the PC Market," *NYT*, 4 December 2004, B1, B3.

29. S. Hamm, "Big Blue's Bold Step into China," *Business Week*, 20 December 2004, 35–36.

30. John Markoff, "Have Supercomputer, Will Travel," *NYT*, 1 November 2004, C1, C4.

31. Scott Thurm, "China's Huawei Halts U.S. Sales amid Cisco Claim," *WSJ*, 7 February 2003, B3; Sameena Ahmad, "We Are the Champions," *The Economist*, "Survey of Business in China," 20 March 2004, 14–15.

32. Scott Thurm, "China's Huawei, 3Com to Form Venture to Compete with Cisco," *WSJ*, 20 March 2003, B5.

33. David Pringle and Nisha Gopalan, "China's Huawei Wins 3G Contract," *WSJ*, 19 February 2003, B2.

34. Christopher Rhoads and Charles Hutzler, "China's Telecom Forays Squeeze Struggling Rivals," *WSJ*, 8 September 2004, A1, A13; Christopher Rhoads and Rebecca Buckman, "A Chinese Telecom Powerhouse Stumbles on Road to the U.S.," *WSJ*, 28 July 2005, A1, A6.

35. Mei Fong, "China Has Much to Gain Going Online," *WSJ*, 5 November 2003,

B4; Mei Fong, "Tom Online to Make Nasdaq Debut," WSJ, 10 March 2004, C16; Kathy Chen, "Now, a New Way Cellphones Are Hot in China," WSJ, 22 September 2003, B1, B3.

36. Fiona Harvey, "China Registers Its Larger Interest in Running of Internet," FT, 26 November 2002, 6.

37. Evan Ramstad, "Chip That Speaks Languages of Asia Levels Playing Field," WSJ, 9 February 2004, B8.

38. Richard Waters, "People Power to Make China Online Capital," FT, 29 November 2004, 7.

39. Yuezhi Zhao, "Caught in the Web: The Public Interest and the Battle for Control of China's Information Superhighway," Info 2.1 (February 2000): 57; Edith M. Lederer, "Official Says China Will Soon Surpass U.S. to Become World's Largest Internet and Information Economy," Associated Press, 25 January 2003; retrieved March 2003, lexis-nexis.com/university/.

40. Waters, "People Power to Make China Online Capital," 7.

41. Joe Leahy and Justine Lau, "Frontier of 'A New Global Contest,'" FT, 1 June 2004, 3.

42. Ben Dolven and Trish Saywell, "China Goes for Private Lessons," WSJ, 6 January 2004, A17.

43. X. Xuehui, "Industrializing Education?" in One China, Many Paths, ed. Chaohua Wang (London: Verso, 2003), 246, 244.

44. "Private Universities May Profit in China," Chronicle of Higher Education, 14 February 2003, A41.

45. Ian Harvey, "The West Must Heed China's Rise in the Global Patent Race," FT, 21 September 2005, 15.

46. Joseph Man Chan and Jack Linhuan Qiu, "China: Media Liberalization under Authoritarianism," in Media Reform: Democratizing the Media, Democratizing the State, ed. Monroe Price, Beata Bouzumilowicz, and Stephen Verhulst (London: Routledge, 2002), 36. A careful analysis of one important urban party newspaper concludes that "the influence of the market is getting stronger and stronger." Zhou He, "Chinese Communist Party Press in a Tug-of-War: A Political-Economy Analysis of the Shenzhen Special Zone Daily," in Power, Money, and Media: Communication Patterns and Bureaucratic Control in Cultural China, ed. C. C. Lee (Chicago: Northwestern University Press 2000), 142. See, in general, Yuezhi Zhao, Media, Market, and Democracy in China (Urbana: University of Illinois Press, 1998); and Yuezhi Zhao, Communication in China: Capitalist Reconstruction and Social Contestation (Lanham, Md.: Rowman and Littlefield, forthcoming).

47. Quoted in Lawrence, "Broadcast News, Chinese Style," 30–31.

48. David Hale and Lyric Hughes Hale, "China Takes Off," Foreign Affairs 82.6 (November–December 2003): 37. The danger posed by "China rising" differs, though, with ideological proclivity and material interest. For some, especially smaller U.S. manufacturers reliant on exports, the danger stems from a Communist dictatorship

that is sapping U.S. economic power through an undervalued currency. For others, the threat comes less from China's artificially manipulating the value of the yuan than from its implication in an ominous Darwinian turn of the globalization process: low-wage manufacturing in China cascades into declining employment standards and deteriorating community values in the United States. The ascent of Wal-Mart, the world's largest private employer and retailer, is cited as paradigmatic; Wal-Mart takes 10 percent of U.S. imports from China. See Jeffrey E. Garten, "Wal-Mart Gives Globalism a Bad Name," *Business Week*, 8 March 2004, 24. In either rendition, the putative danger to U.S. dominance is never far removed.

49. George J. Gilboy, "The Myth behind China's Miracle," *Foreign Affairs* 83.4 (July–August 2004): 38.

50. Stephen Labaton, "Advisor to U.S. Aided Maker of Satellites," NYT, 29 March 2003, C1, C4.

51. Charles Hutzler, "China Launches Manned Spacecraft," WSJ, 15 October 2003, A2; Charles Hutzler, "China's Space Program Is No Small Potatoes," WSJ, 15 September 2003, B1, B3; "China and a New Space Race," *Chronicle of Higher Education*, 26 September 2003, B6.

52. Hutchison Telecommunications and Singapore Technologies Telemedia offered $250 million for a 61.5 percent stake in Global Crossing, an upstart international carrier and one of the bankrupt casualties of the international telecom meltdown. The bidders aimed to pay one cent on the dollar for these network facilities. The transaction required approval from the U.S. Federal Communications Commission and the Committee on Foreign Investment in the United States, which is dominated by executive-branch agencies. Issues of "law enforcement and national security" are typically in the forefront of these reviewing agencies' concerns. See Tom Leithauser, "Global Crossing Buyout Gets European OK; Companies Await Approval from CFIUS, FCC," *Telecommunications Reports Daily*, 17 January 2003, 5–6; Simon Romero, "Hong Kong Company May Alter Deal to Buy Global Crossing," NYT, 1 March 2003, B2.

53. Neil King Jr., "China Urged to Drop Tech Rule," WSJ, 4 March 2004, B4.

54. Jason Dean, "China Considers Venture Fund for Chip Industry," WSJ, 26 November 2004, B2.

55. Federal Communications Commission, "International Bureau Releases 2000 Year-End Circuit Status Report for U.S. Facilities-Based International Carriers Reflecting Steady Growth in Capacity Use," press release, 29 June 2001. Because they do not capture data on circuits provided on a non-carrier basis or by private carriers, FCC figures *under*estimate the growth in private-line circuits.

56. Ibid.

57. Cathy Hsu, "2002 Section 43.82 Circuit Status Data" (Washington, D.C.: Federal Communications Commission, International Bureau, 2003), tables 5 and 6; Federal Communications Commission, "International Bureau Releases 2002 Year-End Circuit Status Report for U.S. Facilities-Based International Carriers," press release, 24 December 2003.

58. Cathy Hsu, "2000 Section 43.82 Circuit Status Data," (Washington, D.C.: Federal Communications Commission International Bureau, 2001), 3.

59. Hsu, "2000 Section 43.82 Circuit Status Data," tables 5 and 6; Hsu, "2002 Section 43.82 Circuit Status Data," tables 5 and 6. Generally, the number of active circuits increased annually through 2000 or 2001 and then declined somewhat, while the proportion of public service to private line service continued to decline or remained stable over this interval.

60. Hsu, "2000 Section 43.82 Circuit Status Data," tables 5 and 6; Hsu, "2002 Section 43.82 Circuit Status Data," tables 5 and 6.

61. Hsu, "2000 Section 43.82 Circuit Status Data," tables 5 and 6; Hsu, "2002 Section 43.82 Circuit Status Data," tables 5 and 6.

62. R. D. Lyons. "China Developing Satellite Links," NYT, 5 January 1973, 4; Cathy Hsu, "1995 Section 43.82 Circuit Status Data" (Washington, D.C.: Federal Communications Commission, International Bureau, 1996), tables 5 and 6.

63. Hsu, "2000 Section 43.82 Circuit Status Data," tables 5 and 6; Hsu, "2002 Section 43.82 Circuit Status Data," tables 5 and 6.

64. Hsu, "2000 Section 43.82 Circuit Status Data," 4; Hsu, "2002 Section 43.82 Circuit Status Data," 4.

65. Dan Schiller, "World Communications in Today's Age of Capital," Emergences 11.1 (2001): 59.

66. Jesse Drucker, "Global Talk Gets Cheaper," WSJ, 11 March 2004, B1, B2; Mei Fong, "The Spam-China Link," WSJ, 19 March 2004, B1, B2; United Nations Commission on Trade and Development, World Investment Report (New York: United Nations, 2004).

67. Rebecca Buckman and Julie Wang, "PCCW Nears Deal to Sell Stake in Core Asset to China Netcom," WSJ, 25 August 2004, B7.

68. Yuezhi Zhao and Dan Schiller, "Dances with Wolves? China's Integration with Digital Capitalism," Info 3.2 (April 2001): 137–51.

69. Yuezhi Zhao, "Transnational Capital, Chinese Capitalism, and China's Semi-Integrated Communication Industries in a Fractured Society," Paper presented at the Rockefeller Foundation Conference Center, Bellagio, Italy, 12–16 May 2003.

70. "China now has a stake in the liberal, rules-based global economic system that the United States worked to establish over the past half-century" (Gilboy, "Myth behind China's Miracle," 33).

71. Justine Lau, "Most Chinese Groups Prefer Not to Go Global," FT, 29 September 2005, 7.

72. For a historical study of this endeavor in its U.S. context that focuses on the neglected but vital contributions made by policy for what we now call intellectual property, see Doron S. Ben-Atar, Trade Secrets: Intellectual Piracy and the Origins of American Industrial Power (New Haven, Conn.: Yale University Press, 2004).

73. Saumitra Chaudhury, "Indian Bourgeoisie and Foreign Capital: A Study of

Congress Policy towards Foreign Capital, 1931–1961," *Social Scientist* (New Delhi) 12.5 (May 1984): 3–22.

74. Gilboy, "Myth behind China's Miracle," 36–37.

75. This point comes through vividly in a recent novel set in the Three Gorges context: Hong Ying, *Peacock Cries at the Three Gorges*, trans, Mark Smith and Henry Zhao (London: Marion Boyars, 2004).

76. Jiang Zemin, "Hold High the Great Banner of Deng Xiaoping: Theory for an All-Round Advancement of the Cause of Building Socialism with Chinese Characteristics into the Twenty-First Century," Report Delivered at the Fifteenth National Congress of the Communist Party of China, 12 September 1997, retrieved 6 February 2006, http://ce.cei.gov.cn/efor/b3100j01.htm.

77. Communist Party of China, Constitution of the Communist Party of China Amended and Adopted at the Sixteenth CPC National Congress, 14 November 2002, retrieved 6 February 2006, www.china.org.cn/english/features/49109.htm.

78. R. Wu, "Making an Impact," *Nature* 428, Supplement (11 March 2004): 206.

79. Xiaozhong Yang, "An Embryonic Nation," *Nature* 428, Supplement (11 March 2004): 211.

80. Ahmad, "We Are the Champions," 14–15.

81. D. Cyranoski, "China Increases Share of Global Scientific Publications," *Nature* 431 (9 September 2004): 116.

82. John Harwood, "Competitive Edge of U.S. Is at Stake in the R&D Arena," WSJ, 17 March 2004, A4; Charlene Barshefsky and Edward Gresser, "Revolutionary China, Complacent America," WSJ, 16 September 2005, A20.

83. Pete Engardio, Aaron Bernstein, and Manjeet Kripalani, "The New Global Job Shift," *Business Week Online*, 3 February 2003, retrieved 6 February 2006, www.businessweek.com/@@hzrhI4UQZFTjcBOA/magazine/content/03 05/b3818001.htm; Matt Murray, "GE's Immelt Starts Renovations on the House That Jack Built," WSJ, 6 February 2003, A1, A6.

84. Daniel Altman, "China: Partner, Rival, or Both?" NYT, 2 March 2003, sec. 3, 1, 11.

85. L. Santini, "Drug Companies Look to China for Cheap R&D," WSJ, 22 November 2004, B1, B4.

86. N. D. Shi, "Shi Guangsheng on Achievements of China's Foreign Trade and Economic Cooperation," January 2003, retrieved 6 February 2006, www.china.org.cn/English/features/48710.htm.

87. Matthew Karnitschnig, "Vaunted German Engineers Face Competition from China," WSJ, 15 July 2004, A1, A8; Chris Buckley, "Let a Thousand Ideas Flower: China Is a New Hotbed of Research," NYT, 13 September 2004, C1, C4.

88. Quoted in Lederer, "Official Says China Will Soon Surpass U.S. to Become World's Largest Internet and Information Economy"; see also Hale and Hale, "China Takes Off," 43–45.

89. Quoted in Lederer, "Official Says China Will Soon Surpass U.S. to Become World's Largest Internet and Information Economy."

90. Hale and Hale, "China Takes Off," 47.

91. United Nations Commission on Trade and Development, *Trade and Development Report 2002* (New York: United Nations, 2002), 154; Shi, "Shi Guangsheng on Achievements of China's Foreign Trade and Economic Cooperation."

92. Peter Wonacott, "U.S. Pursues a Trade Ally in Beijing," WSJ, 18 February 2003, A20, A21; Andrew Browne, "China Drew over $60 Billion in Foreign Investment in 2005," WSJ, 14–15 January 2006, A2.

93. Hale and Hale, "China Takes Off," 38. Total profits earned by foreign-funded companies in China came to around $20 billion in 2000, and although profits remain uneven and volatile, these companies reinvested $12 billion of this total in China. United Nations Commission on Trade and Development, *Trade and Development Report 2002*, 155. According to one report, direct and indirect profits made by U.S. corporate affiliates in China totaled $2.8 billion in 2001—considerably less than their counterparts earned in Mexico ($4.4 billion). A different study found, however, that in 2002, three-quarters of American Chamber of Commerce member corporations active in China claimed to be profitable, "and nearly 40 percent said their margins in China were higher than their global margins." Sameena Ahmad, "Bulls in a China Shop," *The Economist,* "Survey of Business in China," 20 March 2004, 10, 9.

94. Ahmad, "We Are the Champions," 14–15; Owen Brown and Andrew Browne, "China Opens Door for Its Companies to Invest Overseas," WSJ, 13 October 2004, A2.

95. Ahmad, "We Are the Champions," 14–15.

96. Frances Williams, "China 'Set to Join League of Biggest Direct Investors Abroad,'" FT, 5 May 2004, 5.

97. Ahmad, "We Are the Champions," 14–15.

98. Gilboy, "Myth behind China's Miracle," 43.

99. United Nations Commission on Trade and Development, *World Investment Report 2002*, tables 2–8, 56.

100. P. Day, "China's Trade Lifts Neighbors," WSJ, 18 August 2003, A9.

101. United Nations Commission on Trade and Development, *Trade and Development Report 2002*, 162–63.

102. Ahmad, "We Are the Champions," 14–15.

103. United Nations Commission on Trade and Development, *Trade and Development Report 2002*, 163.

104. Ibid., 164.

105. Hale and Hale, "China Takes Off," 47; Howard W. French and Norimitsu Onishi, "Economic Ties Binding Japan to Rival China," NYT, 31 October 2005, A1, A10.

106. Dal Yong Jin, "Political Economy of Communication Industry Reorganization:

Republic of Korea, 1987–2002" (Ph.D. dissertation, University of Illinois at Urbana-Champaign, 2004), 359–60.

107. Minxin Pei, "A Docile China Is Bad for Global Peace," FT, 12 March 2003, 13; "China's Power Play" (editorial), WSJ, 4 August 2005, A12; Peter Wonacott and Neil King Jr., "China Irks U.S. as It Uses Trade to Embellish Newfound Clout," WSJ, 3 October 2005, A1, A14.

108. Robert A. Guth, "Asia to Develop Software to Rely Less on Microsoft," WSJ, 17 November 2003, B4.

109. Michele Yamada, "Asian Countries Seek Windows Alternative," WSJ, 2 September 2003, B10.

110. Ramstad, "Chip That Speaks Languages of Asia Levels Playing Field," B8.

111. G. McCormack, "Remilitarizing Japan," New Left Review 29 (September–October 2004): 29–45; Greg Jaffe and Neil King Jr., "U.S. See Broad China Threat in Asia," WSJ, 20 July 2005, A6.

112. G. York, "Nationalist Fervour Runs Amok," Toronto Globe and Mail, 25 October 2004, retrieved 25 October 2004, www.theglobeandmail.com/servlet/ArticleNews/TPSttory/LAC/20041025/CHINANAT25/International/Idx; Adam Gamble and Takesato Watanabe, A Public Betrayed: An Inside Look at Japanese Media Atrocities and Their Warnings to the West (New York: Regnery, 2004).

113. Kanji Ishibashi and Phred Dvorak, "H-P to Sell PCs Running on Linux in Asian Market," WSJ, 17 March 2004, B5.

114. United Nations Commission on Trade and Development, Trade and Development Report 2002, 155.

115. In 2002, China supplied 18.3 percent ($61.7 billion) of Japan's imports, while the United States accounted for 17 percent ($57.5 billion). Associated Press, "China Becomes Biggest Exporter to Japan," NYT, 19 February 2003, W1.

116. Greg Ip, "Trade Gap Widens to Record Level," WSJ, 21 February 2003, A2.

117. Shi, "Shi Guangsheng on Achievements of China's Foreign Trade and Economic Cooperation"; Craig Karmin, Phred Dvorak, Phillip Day, and Michael R. Sesit, "Japan Pays for Low U.S. Interest Rates," WSJ, 18 March 2004, C1, C2.

118. U.S. Department of the Treasury, "Major Foreign Holders of Treasury Securities 2004," retrieved March 2004, www.treasury.gov/tic/mfh.txt and www.treasury.gov/tuc/mfhhis01.txt.

119. Hale and Hale, "China Takes Off," 50; Craig Karmin and Karen Richardson, "Sliding Dollar's Fate May Be Decided in Asia," WSJ, 20 January 2003, C1, C12.

120. Jill R. Newbold, "Aiding the Enemy: Imposing Liability on U.S. Corporations for Selling China Internet Tools to Restrict Human Rights," Journal of Law, Technology, and Policy 2 (Fall 2003), retrieved 6 February 2006, www.jtp.uiuc.edu/current/newbold.pdf; Bruce Einhorn and Ben Elgin, "The Great Firewall of China," Business Week, 23 January 2006, 32–34.

121. Richard B. Du Boff, "NAFTA and Economic Integration in North America:

Regional or Global?" in *Continental Order? Integrating North America for Cyber-capitalism*, ed. Vincent Mosco and Dan Schiller (Lanham, Md.: Rowman and Littlefield, 2001), 35–63.

122. Robert Brenner, *The Boom and the Bubble: The U.S. in the World Economy* (London: Verso, 2002).

123. For a powerful depiction of the effects on China's peasantry, see L. Changping, "The Crisis in the Countryside," in *One China, Many Paths*, ed. Chaohua Wang (London: Verso, 2003), 198–218.

124. Joel Baglole, "Citibank Takes Risk by Issuing Cards in China," WSJ, 10 March 2004, C1, C2.

125. Karby Leggett and Kathy Chen, "For China, Question of Debt Is Crucial," WSJ, 20 January 2003, A2.

126. James Kynge and Mure Dickie, "Chinese PM Warns of Threat to Rapid Growth," FT, 15 March 2004, 1.

127. Sameena Ahmad, "Behind the Mask," *The Economist*, "Survey of Business in China," 20 March 2004, 3.

128. "China Sets Curbs on Money and Credit Growth," WSJ, 25 March 2004, A15; Matt Pottinger, "China's Premier Lists Concerns about Economy," WSJ, 15 March 2004, A15; R. Glenn Hubbard, "Market Comrades," WSJ, 26 July 2005, A24.

129. Sameena Ahmad, "A Billion Three, but Not for Me," *The Economist*, "Survey of Business in China," 20 March 2004, 6; Peter Wonacott, Joseph B. White, and Norihiko Shirouzu, "Car Companies Jockey for Slice of China Market," WSJ, 8 June 2004, A13, A15; Peter Wonacott, "China's Hot Auto Sales Cool," WSJ, 24 November 2004, B2. Late in 2005, a Chinese official claimed that the auto industry "is facing a grave overproduction situation," estimating that by 2010, China's total annual production capacity will exceed twenty million vehicles, while demand will be limited to nine million. Shai Oster, "China Frets over Auto-Capacity Glut," WSJ, 22 November 2005, A13.

130. Neal E. Boudette, Jathon Sapsford, and Peter Wonacott, "Cars Made in China Are Headed to the West," WSJ, 22 April 2005, B1, B2 (quote); Peter Wonacott, "Global Aims of China's Car Makers Put Existing Ties at Risk," WSJ, 24 August 2004, B1, B3.

131. And, concurrently, a higher incidence of large-scale social protest: such protests were said to be rising by 50 percent a year, according to the Public Security Ministry. Charles Hutzler, "China's Top Cop Wields Nightstick More Subtly Than Beijing Used To," WSJ, 13 March 2003, A11; Zigang Dong, Christina W. Hoven, and Allan Rosenfield, "Lessons from the Past," *Nature* 43 (10 February 2005): 573–74.

132. Jonathan Yardley, "Farmers Being Moved Aside by China Real Estate Boom," NYT, 8 December 2004, A1, A10.

133. T. C. Tso, "Agriculture and the Future," *Nature* 428, Supplement (11 March 2004): 215.

134. J. Kahn "China Crushes Peasant Protest, Turning 3 Friends into Enemies," NYT, 13 October 2004, A1, A8. Yuezhi Zhao, *Communication in China*, provides a compelling account.

135. Yang Lian, "Dark Side of the Chinese Moon," *New Left Review* 32 (March–April 2005): 139; Kathy Chen, "Beijing's Blueprint to Tackle Gap between Rich, Poor," WSJ, 30 September 2005, A9.

136. Ahmad, "Behind the Mask," 3.

Index

DAN SCHILLER is a professor in the Graduate School of Library and Information Science and the Institute of Communications Research at the University of Illinois at Urbana-Champaign. He is a communication historian whose interests center on telecommunications history, and on the role of cultural production in the socio-economic development of the market system. His books include *Digital Capitalism: Networking the Global Market System; Theorizing Communication: A Historical Reckoning; Telematics and Government;* and *Objectivity and the News: The Public and the Rise of Commercial Journalism.*

The University of Illinois Press
is a founding member of the
Association of American University Presses.

Composed in 10.5/13 Adobe Minion
with Meta display
by Celia Shapland
for the University of Illinois Press
Manufactured by Thomson-Shore, Inc.

University of Illinois Press
1325 South Oak Street
Champaign, IL 61820-6903
www.press.uillinois.edu

DATE DUE

JUN 1 7 2008			

Demco No. 62-0549